S. Hrg. 106–352

BIOTERRORISM—DOMESTIC WEAPONS OF MASS DESTRUCTION

JOINT HEARING

BEFORE THE

SUBCOMMITTEE ON

LABOR, HEALTH AND HUMAN SERVICES, AND EDUCATION, AND RELATED AGENCIES

COMMITTEE ON APPROPRIATIONS

AND THE

COMMITTEE ON VETERANS' AFFAIRS

UNITED STATES SENATE

ONE HUNDRED SIXTH CONGRESS

FIRST SESSION

SPECIAL HEARING

Printed for the use of the Committees on Appropriations and Veterans' Affairs

Available via the World Wide Web: http://www.access.gpo.gov/congress/senate

U.S. GOVERNMENT PRINTING OFFICE

57–119 cc WASHINGTON : 2000

For sale by the U.S. Government Printing Office
Superintendent of Documents, Congressional Sales Office, Washington, DC 20402
ISBN 0–16–060194–0

COMMITTEE ON APPROPRIATIONS

TED STEVENS, Alaska, *Chairman*

THAD COCHRAN, Mississippi
ARLEN SPECTER, Pennsylvania
PETE V. DOMENICI, New Mexico
CHRISTOPHER S. BOND, Missouri
SLADE GORTON, Washington
MITCH McCONNELL, Kentucky
CONRAD BURNS, Montana
RICHARD C. SHELBY, Alabama
JUDD GREGG, New Hampshire
ROBERT F. BENNETT, Utah
BEN NIGHTHORSE CAMPBELL, Colorado
LARRY CRAIG, Idaho
KAY BAILEY HUTCHISON, Texas
JON KYL, Arizona

ROBERT C. BYRD, West Virginia
DANIEL K. INOUYE, Hawaii
ERNEST F. HOLLINGS, South Carolina
PATRICK J. LEAHY, Vermont
FRANK R. LAUTENBERG, New Jersey
TOM HARKIN, Iowa
BARBARA A. MIKULSKI, Maryland
HARRY REID, Nevada
HERB KOHL, Wisconsin
PATTY MURRAY, Washington
BYRON L. DORGAN, North Dakota
DIANNE FEINSTEIN, California
RICHARD J. DURBIN, Illinois

STEVEN J. CORTESE, *Staff Director*
LISA SUTHERLAND, *Deputy Staff Director*
JAMES H. ENGLISH, *Minority Staff Director*

SUBCOMMITTEE ON LABOR, HEALTH AND HUMAN SERVICES, AND EDUCATION, AND RELATED AGENCIES

ARLEN SPECTER, Pennsylvania, *Chairman*

THAD COCHRAN, Mississippi
SLADE GORTON, Washington
JUDD GREGG, New Hampshire
LARRY CRAIG, Idaho
KAY BAILEY HUTCHISON, Texas
TED STEVENS, Alaska
JON KYL, Arizona

TOM HARKIN, Iowa
ERNEST F. HOLLINGS, South Carolina
DANIEL K. INOUYE, Hawaii
HARRY REID, Nevada
HERB KOHL, Wisconsin
PATTY MURRAY, Washington
DIANNE FEINSTEIN, California
ROBERT C. BYRD, West Virginia
(Ex officio)

Professional Staff
BETTILOU TAYLOR
MARY DIETRICH
JIM SOURWINE
AURA DUNN
ELLEN MURRAY *(Minority)*

Administrative Support
KEVIN JOHNSON
CAROLE GEAGLEY *(Minority)*

COMMITTEE ON VETERANS' AFFAIRS

ARLEN SPECTER, Pennsylvania, *Chairman*

STROM THURMOND, South Carolina
FRANK H. MURKOWSKI, Alaska
JAMES M. JEFFORDS, Vermont
BEN NIGHTHORSE CAMPBELL, Colorado
LARRY E. CRAIG, Idaho
TIM HUTCHINSON, Arkansas

JOHN D. ROCKEFELLER IV, West Virginia
BOB GRAHAM, Forida
DANIEL K. AKAKA, Hawaii
PAUL WELLSTONE, Minnesota
PATTY MURRAY, Washington

CHARLES BATTAGLIA, *Staff Director*
JIM GOTTLIEB, *Minority Chief Counsel/Staff Director*

CONTENTS

BIOTERRORISM—DOMESTIC WEAPONS OF MASS DESTRUCTION

TUESDAY, MARCH 16, 1999

U.S. SENATE, SUBCOMMITTEE ON LABOR, HEALTH AND HUMAN SERVICES, AND EDUCATION, AND RELATED AGENCIES, COMMITTEE ON APPROPRIATIONS, AND COMMITTEE ON VETERANS' AFFAIRS,
Washington, DC.

The subcommittee and Committee met jointly at 9:33 a.m., in room SD–106, Dirksen Senate Office Building, Hon. Arlen Specter (chairman) presiding.

Present: Senators Specter, Stevens, Kyl, Campbell, and Rockefeller.

DEPARTMENT OF VETERANS' AFFAIRS

STATEMENT OF KENNETH W. KIZER, M.D., UNDER SECRETARY FOR HEALTH

DEPARTMENT OF HEALTH AND HUMAN SERVICES

STATEMENT OF MARGARET A. HAMBURG, M.D., ASSISTANT SECRETARY FOR PLANNING AND EVALUATION

GENERAL ACCOUNTING OFFICE

STATEMENT OF HENRY L. HINTON, JR., ASSISTANT COMPTROLLER GENERAL FOR NATIONAL SECURITY AND INTERNATIONAL AFFAIRS

OPENING STATEMENT OF SENATOR ARLEN SPECTER

Senator SPECTER. Good morning, ladies and gentlemen. This joint hearing of the Veterans' Affairs Committee and the Appropriations Subcommittee on Labor, Health and Human Services will now proceed.

Our subject this morning is domestic weapons of mass destruction. The issue of weapons of mass destruction is one of overwhelming importance in America today and, for that matter, throughout the world. As we speak, the Senate is considering legislation which would establish a national missile defense for rogue terrorist nations.

The question of biological warfare and chemical warfare has been one of enormous importance as it applies to the domestic scene. Congress has appropriated very large sums of money to the FBI on counterterrorism. There is a commission now at work to deal with the governmental organization on weapons of mass destruction.

Legislation was inserted into the 1996 defense authorization bill, at a time when I chaired the Intelligence Committee, to take a fresh look at the 96 separate agencies which deal with weapons of mass destruction. High on the list is the question of what happens domestically, where responsibility has been lodged with the Department of Health and Human Services and the Department of Justice and the Veterans Administration.

Today, we are going to make inquiries into what line of preparedness there is at the present time and what ought to be done. This hearing is held in coordination with the work of the Commission on Weapons of Mass Destruction chaired by former CIA Director John Deutch to try to get some insights to see what more ought to be done or how our organization ought to be structured to take appropriate stock of this very, very serious problem.

Today we have six distinguished witnesses and our lead witness is Dr. Kenneth W. Kizer, Under Secretary for Health in the U.S. Department of Veterans Affairs. As the chief executive officer of the VA Administration and the highest ranking physician in the Federal Government, Dr. Kizer oversees the 173 VA hospitals and administers a medical budget of $17 billion, not enough, but a starting point.

Dr. Kizer, let me put a pointed, specific question to you as to the role of the Veterans Administration in dealing with the issue of weapons of mass destruction on the domestic level.

Before I do, we have Senator Campbell with us. Senator Campbell, would you care to make an opening statement?

OPENING STATEMENT OF SENATOR BEN NIGHTHORSE CAMPBELL

Senator CAMPBELL. Thanks, Mr. Chairman. I have a conflict, so I will have to run in a few minutes. But I am very interested in this hearing.

Coincidentally, just a few months ago we had three families in Colorado Springs, which is our second largest city in Colorado, that received letters and packages with warnings that the package contained anthrax. I cannot imagine anybody sending a package where they would label it on the outside, but that did happen. It did throw a big scare into our local response teams, our HAZMAT teams out there.

As we move along, I know it will not be strictly in the purview of this Committee, but I would like to—I would hope that as you do move along we do some exploration with other committees about how we are going to interact with the Federal Government and local teams.

Some firemen came in this morning, in fact, and they were telling this about these packages last fall. The firemen that were in this morning, they described themselves as test rats. They said they get so little help from the Federal Government that when there is any kind of a terrorist threat, when they have to go out they go out totally unequipped with knowledge or equipment and it is kind of by hook or crook. If some die they must have the wrong equipment. If they manage to survive the thing, then they must have the right equipment.

That is a pretty tragic way to respond to any kind of a threat of some of these biological weapons. So as we do move along I would hope this Committee would look into some interaction with other committees on seeing if we cannot provide some Federal assistance to those local teams.

Thank you, Mr. Chairman.

Senator SPECTER. Thank you very much, Senator Campbell.

Our distinguished ranking member, Senator Rockefeller, for an opening statement.

OPENING STATEMENT OF SENATOR JOHN D. ROCKEFELLER, IV

Senator ROCKEFELLER. Thank you, Mr. Chairman, very much.

I do not know who Daniel Greenberg is, but he probably ought to move to Oregon as quickly as possible, because this has to be one of the most ridiculous articles I have ever read in my life—his op-ed piece in this morning's Washington Post—in which he blithely dismisses the subject that a lot of people worry about, in a child-like way. He may be a scientist. I have no idea. But I think he should move to Oregon.

I think that it is a serious problem and I think that——

Senator SPECTER. How far is Oregon from West Virginia, Senator?

Senator ROCKEFELLER. Oregon is a long way from West Virginia, a long way from West Virginia, and may it ever remain so.

I think the potential for exaggeration, which is, of course, what he emphasizes—that this concern is all exaggerated—obviously, in anything of this sort, the potential for exaggeration is high because it is speculative. On the other hand, it is extremely real. It has always occurred to me, just in the nature of common sense, that entrepreneurial or ideological terrorism, biological or chemical, is a part of what we are facing in this country and is as great a threat or greater than the threat that the Soviet Union posed to us.

Ebola, smallpox, anthrax, all of these things, whether they are imagined or not, are there and have the potential to bring unspeakable horrors upon this country.

So I am glad that the administration has taken this threat seriously, or appears to be. It is an enormously complicated matter, the coordination of a response. I think in the Department of Defense I counted 19 separate divisions which deal with this, and I think in HHS I counted 10. So that is two agencies, 29 defense groups.

So, we are undertaking, as Senator Campbell has just indicated, a very difficult thing, in which States and their agencies have to try to cooperate on this. It is fascinating to me and very depressing that only about half of the public health facilities in this country even have the computers with which to try to do the best they can to communicate with CDC, which by definition, therefore, would be an impossibility. We are way, way, way behind in our efforts to deal with bioterrorism in whatever form.

I think generally people would agree that one of the biggest challenges is fortifying our medical system, including improving our laboratory capability and vaccine stockpiling. Most U.S. hospitals, including VA hospitals, are not necessarily prepared to treat patients contaminated by chemical or biological agents, and we can discuss that.

I think our medical system must be merged with grassroots public health offices, at least half of which are virtually unprepared to deal with anything of this sort, even to communicate about it.

Our challenge is to build upon the existing public health system and infrastructure to vastly improve the medical response, and, of course, that is going to be very difficult. In preparing for an act of bioterrorism, we are going to have to spend plenty of money.

I do not want to make a long statement, but I want to make four points, four important points for me:

First, how are we going to ensure that our emergency room doctors and other frontline health professionals—assuming that they themselves were not affected by some kind of attack—will be trained to recognize and treat germ warfare diseases?

In other words, somebody walks in with a biological attack, which may not show up for a week or two, and has respiratory problems, coughing, sneezing, and other things. Is that going to be diagnosed? Are our medical people prepared to diagnose what may appear to be something quite different and is not, but in the meantime, is communicable?

Second, given the vast differences between dealing with the release of chemical weapons versus biological weapons, how can we simultaneously and successfully pursue remedies to each of these two?

Third, shortages of vaccines pose a potentially very dangerous situation, and that, of course, is especially true in the case of smallpox. I thought that counterposing Daniel Greenberg's dribblings in this morning's Washington Post was a slightly more interesting article in the New York Times, which said that perhaps resurrection of the study of smallpox and things of that sort could potentially be very useful.

Fourth, what can be done to ensure that there is close coordination between, as Senator Campbell indicated, major government players at all levels? We live with this in the Canaan Valley in West Virginia because we have the highest concentration of chemicals in that area of any place in the country, and we are constantly on alert because we have high populations surrounding very, very difficult chemical situations.

We are very practiced at it and very good at it, and people come from all around the country to look at how we prepare for this. But of course, that kind of potential disaster is nothing like what we are talking about here.

One more idea, Mr. Chairman, and that is, when you have such a complex and such a high profile issue, involving so many, there is often need for a strong national civilian coordination of the many competing agencies, coordination that continues from one administration to the next, that is not Republican or Democrat, is not Bush or Clinton. It just goes forward.

I have a lot of concerns that we do not have that situation in place now. We have, I think, a very good person who is in charge of doing it for this administration. Does that mechanism continue or is there a better mechanism? One solution would be creation of a domestic terrorism response advisory board, and I would like to have the views of witnesses on that, as a permanent entity to provide overall national guidance. This would be fashioned on the model of the Presidential Foreign Intelligence Advisory Board, which I think people think has worked fairly well.

I thank the chairman.

PREPARED STATEMENT OF SENATOR DIANNE FEINSTEIN

Dianne Feinstein asks that I insert a statement in the record, and who am I to say no.

Senator SPECTER. Without objection, it will be submitted.

[The statement follows:]

PREPARED STATEMENT OF SENATOR DIANNE FEINSTEIN

Thank you, Chairman Specter and Ranking Minority Member Harkin for holding today's hearing on biological terrorism.

I look forward to hearing today's testimony because I believe many of us are very concerned about how prepared our nation might be for these biological threats. Any threat of terrorism is a great cause of concern. Biological terrorism has its own unique dimensions because biological agents can do great harm with small quantities; they can be relatively easily concealed and delivered; they can spread quickly; and their symptoms can take days to develop and are difficult for health care workers to diagnose.

As a former mayor who had to worry about prompt emergency responses to crises of all kinds—from accidents to fires, from AIDS to earthquakes—I am particularly concerned about the readiness of our traditional emergency response teams to deal with biological exposures.

No one is immune from this form of terrorism. These agents can be put into our water, our food, our heating and cooling systems. In Washington, there have been anthrax scares at the State Department, the Washington Post, NBC News, the Old Executive Office Building, and here in the Congress.

But these scares are not confined to Washington. There has been a virtual rash of them in California: in schools, a nightclub, an office building, a courthouse and a department store. The Washington Post reported on January 11, with a dateline of Los Angeles:

> A wave of hoaxes involving the lethal bacteria is spreading across Southern California and turning up in states nationwide. It is a fad so alarming, so costly and so confounding to police and public health officials that some almost sound wistful for the days when they had to contend only with phony bomb scares. Since late last year, nearly two dozen anthrax threats have been reported just in greater Los Angeles.

Most experts caution that we are not prepared. For example, at a conference last year, Dr. Donald Henderson, an expert from Johns Hopkins University who will testify today, said:

> The U.S. lacks the infrastructure, planning and fundin . . . U.S. efforts to deal with biological weapons used against a civilian population are only two years old. These actions are only marginally funded and marginally supported.

Similarly, the December 30, 1998 Los Angeles Times reported:

> Recent anthrax threats in Southern California dramatically underscore the lack of a comprehensive national plan to guide health agencies responding to biological or chemical terrorism. . . .

I welcome the budget proposal from the Administration to improve our ability to respond to biological terrorism. I look forward to working with you, Mr. Chairman, and the Administration to put adequate resources into this effort and to reassure the American people that we can anticipate, detect and respond effectively in hopes that a strong defense can deter these egregious acts.

Senator SPECTER. Senator Kyl, do you care to make an opening statement?

Senator KYL. No, Mr. Chairman. Thank you.

SUMMARY STATEMENT OF DR. KENNETH W. KIZER

Senator SPECTER. We had just begun to pose a question to you, Dr. Kizer. I asked the question instead of calling on you for an opening statement because I am advised that the Office of Management and Budget has not cleared your statement. So let us begin with a generalized description of the role which the Veterans Administration has pursuant to Presidential Decision Directive 62, issued by President Clinton on May 18, 1998, which ordered Federal agencies to expand steps to protect against biological and other conventional domestic attacks.

Dr. KIZER. Thank you, Mr. Chairman, and good morning, members of the Committee. I am pleased to appear before you this morning to talk about both current and potential roles for the Department of Veterans Affairs in Federal emergency management in general and with regard to weapons of mass destruction in particular.

As you noted already, I do not have a cleared formal statement, so what I would like to offer here at the outset is both a general response as well as a personal reflection based on some 25 years, or so, of experience in disaster planning and emergency management. That includes specific experience as a fireman, as a naval officer involved particularly in some of these areas, as an emergency physician, as a medical toxicologist, as a public health official, and as one who has managed a number of infectious disease emergencies or epidemics, as well as the largest pesticide poisoning epidemic in North American history.

In brief, I believe this country is woefully unprepared for a terrorist incident involving weapons of mass destruction, and I believe, regretfully, that this lack of preparedness translates, or would translate, should such occur today, into unnecessary loss of life and suffering. I think, however, there is much that could be done in the near term to better prepare us for such an inevitable event.

I would note for the record that, at least in my mind, just as there is no doubt that floods and earthquakes and hurricanes and tornadoes and other such natural disasters will strike the United States in the future, there is no question in my mind that a terrorist event involving weapons of mass destruction will occur in the United States, and the real question is really only a matter of when and where such will occur and to what extent we are prepared.

Having offered that perspective, in more specific response to your question, I would note that the Department of Veterans Affairs today provides a supportive role to other agencies in Federal emergency management plans. We also have a specific role in managing some pharmaceutical caches that might be needed in the event of a relevant chemical incident. Let me leave it at that for the moment.

ADDITIONAL COMMITTEE QUESTIONS

Senator SPECTER. All right, thank you very much, Dr. Kizer. We will return to you during the more extended question and answer session.

There will be some additional questions which will be submitted for your response in the record.

[The following questions were not asked at the hearing, but were submitted to the Department for response subsequent to the hearing:]

DRAFT RESPONSES OF HON. KENNETH W. KIZER TO QUESTIONS SUBMITTED BY SENATOR ARLEN SPECTOR, CHAIRMAN, COMMITTEE ON VETERANS' AFFAIRS

Question. What is VA's current role in assisting localities in preparing for, and participating in, a medical response to WMD incidents? Do you believe that VA resources have been adequately incorporated into the overall Federal plan to assist localities? What areas do you envision VA playing a larger role?

Answer. The current WMD preparedness role of VA was assigned by PDD–62, which tasked PHS, working with VA, to ensure adequate stockpiles of antidotes and other necessary pharmaceuticals nationwide and to train medical personnel in NDMS hospitals. PHS has provided funding for four pharmaceutical caches, and there is a provision in the HHS fiscal year 2000 budget for up to $1 million to be transferred to VA for NDMS training. Also, under the Federal Response Plan, Emergency Support Function No. 8, Health and Medical, VA is prepared to respond to WMD incidents as well as any other local and regional disasters.

VA's potential for an expanded WMD role has not received appropriate attention. VA, with its institutional capabilities and infrastructure, could be assigned a larger role in this regard.

VA, with the nation's largest fully integrated healthcare system, can uniquely contribute to WMD initiatives in the following areas:

—Use of pharmaceutical infrastructure for procurement, management, and storage.
—Training of health professionals
—Medical and clinical capability
—Coordination with state and local agencies
—Provision of logistical and other support for other federal agencies
—Research

Question. What funding is VA receiving for its WMD related activities? What is the source of this funding? How is this funding expressed in terms of full-time equivalent (FTE) employees? Is funding increasing for VA involvement in WMD activities? In what areas do you see a funding shortfall?

Answer. The Department of Veterans Affairs in compliance with Title 38 United States Code 530 submitted its first "Annual Report on Program and Expenditures for Domestic Response to Weapons of Mass Destruction" to the Committee on Veterans Affairs. This report indicated the following:

Requirement	Fiscal years— 1998	1999	2000
Funds expended or committed	$6,100	$41,320	[1] $1,056,842
Full-time Equivalent Employee (FTEE)	1.15	2.29	2.29
Source of Funds	([2])	([2])	([3])
Anticipated Funds	N/A	N/A	N/A
Anticipated Personnel Shortfall	N/A	N/A	N/A

[1] The Office of Management and Budget (OMB) has approved $1,000,000 for the Department of Health and Human Services, Public Health Service (PHS) to fund a WMD training program with VA designated as the lead training organization. PHS has the discretion to reimburse VA up to $1,000,000 for expenses incurred in conducting a WMD training program for non-federal hospitals participating in the National Disaster Medical System.

[2] Veterans Health Administration.

[3] Veterans Health Administration/Public Health Service.

Question. I understand that VA already collaborates with local hospitals and emergency medical personnel through its involvement in the National Disaster Medical System (NDMS). Explain VA's involvement in NDMS as it relates to planning and preparing for a WMD incident. Is this existing structure being expanded upon to provide more in-depth training and planning services?

Answer. Under PDD–62, VA is preparing to provide WIND response training to NDMS hospitals utilizing the 40 VA Area Emergency Managers (AEMs) located throughout the nation's major population centers. These AEMs are already responsible for the maintenance of 1,545 MOUs with NDMS hospitals that commit staffed acute care beds to the system during a disaster, and regularly participate with local, regional, and state emergency medical planners in the coordination for mutual support, training, and exercises. Some AEMs have proactively initiated VA medical center and community training for WMD. The Emergency Management Strategic Healthcare Group (EMSHG) has supported pre-positioned assets for "special events" such as the 1996 Summer Olympics, the Economic Summit of the Eight, the 1996 Republican and Democratic Conventions, the President's State of the Union addresses, and the recent Papal Visit. Before the Papal visit, EMSHG conducted a seminar in St. Louis on WMD patient reception and treatment for the civilian hospitals. VA will also participate in support of the NATO 50 meeting here in Washington in April. VA medical centers and the AEMs have become part of the local health care systems, creating a solid base upon which VA can build an enhanced role specifically related to WMD.

VHA's existing emergency management infrastructure can be expanded to accommodate new roles if additional resources are provided, in order to ensure that our first and primary mission of caring for veterans is not compromised.

Question. In a joint VA, DOD, HHS, and FEMA report to Congress on the Federal role in preparing for a medical response to a WMD incident delivered in July 1998, it is recommended that VA identify the medical shortfalls in the NDMS and coordinate the actions of NDMS partners to address the shortfalls. What shortfalls have you found? How are you coordinating with the NDMS partners to address these shortfalls?

Answer. These recommendations of the report to Congress were preceded by a recommendation, deleted in the final report, to accomplish a nationwide assessment, including decontamination and quarantine requirements, of the medical system's capability to receive and treat WMD casualties, to include NDMS hospitals. This initiative was also included in VA's initial fiscal year 2000 budget request, but deleted in the passback. No other agency has requested or funded this recommendation. Consequently, no assessment has been made.

Question. I recently sent letters to 19 mayors across the country inquiring about any role they envision VA playing in the event of a WMD incident. VA was mentioned, by some, as being a possible contributor in the areas of pharmaceutical stockpiling, incident-site casualty treatment, and post-incident casualty treatment. Furthermore, in testimony before the Labor-HHS Appropriations Subcommittee, testimony from an International Association of Fire Chiefs representative urged Congress to consider VA for an important role in these same matters. Do you agree with these assessments?

Answer. Yes. I believe VA has significant untapped capabilities that could support the federal government's capability to respond to WMD incidents. VA facilities are geographically dispersed in essentially all of the nation's major metropolitan areas. We have long-established relationships with 85 percent of the nation's medical schools and some 1,200 universities and colleges overall. Because we are part of local health care systems, we have established relationships with the local emergency medical services systems. We already participate in much of the state and local emergency planning activities, as well as coordinate training programs and exercises. In a word, there already exists a solid base upon which VA can build an enhanced role specifically related to WMD.

As I noted earlier, VA manages four pharmaceutical caches for the NMRTs, but beyond this, VHA's infrastructure certainly lends itself to increasing such support in terms of procurement, storage, and management of pharmaceuticals at multiple sites for ready distribution wherever needed. In fact, in recognition of the key role a VA facility can play to address WMD preparedness planning, the Office of the Attending Physician of the U.S. Capitol signed a Memorandum of Understanding with VA in September 1998 to provide for the procurement, maintenance, and storage of a customized WMD cache, to be kept at the Washington VAMC (which also is one of the NMRT sites).

In this same vein, Public Law 104–201 directs and funds HHS to develop 125 local emergency medical systems, or Metropolitan Medical Strike Teams (MMSTs) in selected cities across the United States. Currently, 22 MMSTs exist in various stages of implementation. It has become apparent by the requests for support to VA's PBM/SHG and EMSHG that the strike teams do not have the infrastructure and training in place to facilitate the purchase, maintenance, and distribution of the WMD-related medical caches.

Moreover, I note that while VA's role and potential to support the national response to WMD has not been widely recognized within the federal government, I am advised that in cities where VA facilities have routinely been actively involved in local discussions, local HHS-led Disaster Medical Assistance Teams (DMATs) typically view VA as an important resource for training and logistical assistance.

Thus, I believe that VA's role could be enhanced, at minimum, to formally and proactively provide expertise and service to the expanding network of MMSTs and

DMATs, rather than as a "resource of last resort." And beyond assistance with matters related to pharmaceuticals, these same VA assets could be drawn upon to improve the federal government's preparedness for WMD incidents in five additional areas—i.e., in terms of (1) training; (2) medical and clinical capability; (3) coordination with state and local agencies; (4) provision of logistical and other support for other federal agencies; and (5) research.

Question. Currently, HHS provides VA money to purchase and store pharmaceutical stockpiles for use by specialized teams responding to WMD attacks. I note that these stockpiles are contained at only four sites: Washington, DC; Los Angeles; Denver; and Winston-Salem. Why only four sites? Shouldn't there be additional stockpiles available to prevent delays in delivering necessary treatment? What is the plan to collect and administer pharmaceuticals in a coordinated and timely fashion? Is there sufficient quantity of appropriate antibiotics and vaccines at these sites?

Answer. The four current sites are to support the four National Medical Response Teams (NMRTs) that have been developed by the Department of Health and Human Services (HHS) for response to a chemical/biological incident. Except for the team in Washington, DC, these are deployable teams and, while they would be available for a local response, have been developed primarily for a response to a WMD event occurring in another city. As such, the pharmaceutical supplies that are stored by VA in these locations would be deployed with the teams. VA has agreed to deliver the pharmaceutical cache to the respective NMRT within two hours of notification.

In my opinion, there should be additional stockpiles, and I understand that HHS is placing stockpiles in major cities, especially where the Metropolitan Medical Strike Teams (MMSTs) are being developed. VA has not been requested to support HHS in this effort. In my view, at a minimum, every major city needs to have sufficient quantities of pharmaceuticals to meet local requirements, given their population at risk, especially for first responders to include hospital emergency rooms that will be providing initial care to these victims. This is especially true for a chemical attack where the window of opportunity for providing life-saving or life-sustaining care is very small. The immediate effects of the majority of these agents are such that there will be insufficient time for these supplies to be delivered from remote locations. I should add that these concerns extend to not only pharmaceutical supplies, but also equipment, such as respirators required for care of hospitalized victims.

To our knowledge, other than that established by VA for the caches to support the NMRTs, there is no plan in effect to administer and collect necessary pharmaceuticals in a timely manner that may be placed in other cities. For the four caches directly managed by VA, there is a centralized management system for ordering, shipping, storing, transport, inventory control, inspection, update and rotation of potency dated pharmaceutical items and medical supplies.

VA did not have a role in determining the items, nor the quantities, which are contained in the caches. The requirements were provided to VA from HHS. There is a very limited amount of antibiotics and no vaccines contained in these caches.

Question. New York City's Chem-Bio Handbook recommends that, for post exposure treatment to anthrax, an individual receive a four-week supply of 500 mg ciprofloxacin or 100 mg doxycycline, taken twice daily, followed by an anthrax vaccine immunization. I note that at each of the four stockpile caches, there is only enough ciprofloxacin and doxycycline to provide a four-week supply to 80 people, and no mention of any anthrax vaccine. How do you explain such a limited supply of these antibiotics given what may be an overwhelming demand for them in the event of an anthrax attack?

Answer. VA did not play a part in the development of the caches. These requirements were provided from HHS to VA. However, it should be recognized that the caches were developed not to provide for, or augment, an individual city's requirements over an extended period of time, but to be used for an immediate response to a terrorist event by the respective NMRT. Follow-up treatment and maintenance medications would have to come from another source.

RESPONSES OF HON. KENNETH W. KIZER TO QUESTIONS SUBMITTED BY SENATOR ROBERT C. BYRD, COMMITTEE ON APPROPRIATIONS

Question. Dr. Kizer, has there been an increase in funding for the West Virginia Emergency Management Operation according to the Weapons of Mass Destruction (WMD) threat? Does the Veterans Health Administration (VHA) have plans to reduce the number of full-time-equivalent employees associated with this effort for the West Virginia Emergency Management Operation?

Answer. No additional funds have been provided for the West Virginia Emergency Management operation for its functions relating to Weapons of Mass Destruction threats. Currently, we have no plans to reduce the level of employment at this center. However, a new Chief Consultant has recently been hired for this program, and she will be reviewing current program needs and VHA requirements and a recent VA Inspector General (VAOIG) report on our Emergency Management Strategic Healthcare Group (headquartered at Martinsburg, West Virginia.) This review may result in restructuring recommendations to better meet anticipated VA mission requirements.

Question. Dr. Kizer, will the VHA allocate more funding within the current budget to improve the internal medical emergency management operation? Can it be assumed that more funding will be required beyond the current operational levels? If so, how much funding is required to make the operation truly viable?

Answer. The VHA allocation for this program is currently $7,046,000. This level of funding allows VHA to fulfill current emergency management responsibilities.

Question. Dr. Kizer, I am seriously concerned about reports that have reached me which indicate that the West Virginia Emergency Management Operations, which is located on the campus of the Martinsburg VA Medical Center, is being slated for cuts at a time when many believe that such programs should receive increased emphasis. What can you tell me about this, Dr. Kizer? Please elaborate for the record.

Answer. Again, a new Chief Consultant has just been hired to review program needs and VHA requirements and the VAOIG report mentioned above, and she will recommend an appropriate organizational structure. The Chief Consultant will continue to be based at the Martinsburg VA Medical Center location, which is the headquarters operation for VHA's national emergency preparedness efforts.

SUMMARY STATEMENT OF MARGARET A. HAMBURG, M.D.

Senator SPECTER. Our next witness is Dr. Margaret A. Hamburg, Assistant Secretary for Planning and Evaluation at the Department of Health and Human Services, principal adviser to the Secretary on policy development and coordination and implementation, including the HHS effort on bioterrorism.

We welcome you, Madam Secretary, and look forward to your testimony.

Dr. HAMBURG. Thank you, Mr. Chairman and members of the Committee, for the opportunity to testify. The Department welcomes your interest in our efforts to address the threat of bioterrorism.

I am joined by colleagues who have responsibility for implementing various parts of our initiative. After I briefly outline the overall strategic approach, we would be pleased to answer questions.

Bioterrorism presents a special set of challenges. Unlike a bomb or a discrete chemical exposure, a terrorist incident involving a biological agent may not be detected or even suspected until people begin to present with serious illness. This may occur at considerable and varying remove from the site of initial exposure, both in terms of onset of disease and geographic location. Spreading circles of infectious disease can significantly extend the damage.

In addition, many of the potential biological agents cause diseases not commonly seen here or routinely dealt with by our medical system. Hence, the population generally has little or no immunity, medical providers are not familiar with the diagnosis and treatment, and scientific research has not been a major focus.

For these reasons, the sound strategy for addressing bioterrorism will be quite different from one that targets the other types of terrorist acts. Our initiative addresses five distinct but related areas: deterrence of biological terrorism; strengthening the public health infrastructure for disease surveillance; medical and public health response; development of a national pharmaceutical stockpile; and research and development.

First, deterrence. The CDC has the responsibility to regulate the shipment of certain hazardous biological organisms and toxins. Regulations require that all facilities sending or receiving shipments of select agents register with the CDC, maintain records of such transfers, and otherwise document their compliance. CDC also fosters safe design and secure operation of laboratories that handle select agents.

If an act of bioterrorism occurs, rapid detection and response rests critically on the existence of a robust infrastructure for public health surveillance. How quickly an exposure is detected, analyzed, understood and addressed will hold dramatic implications for the extent and severity of disease, suffering, and societal disruption.

CDC is working to upgrade public health capacity at every level to address the threat of bioterrorism. The emphasis areas are preparedness, planning by State and local health departments, and education and training of the medical community, improved reporting of cases of unusual illness or suspicious patterns of disease, epidemiological analysis of outbreaks to identify the source and mode of transmission, laboratory identification and characterization of the agents involved, and efficient communications among all players.

Much of the initial responsibility for effective response to a bioterrorism attack rests with local governments, with expanding support from State and Federal agencies. Health systems almost inevitably will be called on to provide such critical tasks as mass patient care, mass immunization or prophylactic drug treatment, mass fatality management, infection control, and decontamination of the environment.

PDD–62 designates HHS as the lead Federal agency to plan and prepare for a national response to medical emergencies arising from the terrorist use of weapons of mass destruction. Our Office of Emergency Preparedness works closely with other agencies to ensure that plans for managing these medical emergencies are well integrated with other emergency response systems.

In addition, OEP contracts with local governments for the creation of metropolitan medical response systems to focus on the chem-bio threat. There are 27 municipalities already on board and we hope to bring the total up to 47 this year and 67 in fiscal year 2000. OEP is also working to strengthen its four national medical response teams and the national disaster medical system overall.

A bioterrorism incident would likely require rapid access to quantities of pharmaceuticals that would not be readily available in any given location and argues strongly for the creation of a national stockpile for civilian use. The CDC has been charged with this activity and the initial focus will be on acquiring antibiotics for treating anthrax, plague, and tularemia, botulinum antitoxin, enhancing the utility of the existing supply of smallpox vaccine, and developing a cache of drugs and equipment for countering chemical attacks. As threats shift and as R and D yields new products, we would expect to modify the stockpile contents as appropriate.

Clearly, our ability to detect and counter bioterrorism depends greatly on medical science and technology. The NIH is reinvigorating its research on infectious organisms likely to be used in terrorist acts, including pathogenesis, immune responses, and genomics. In turn, this work will facilitate development of new rapid diagnostic methods, new antiviral and-or antibiotic therapies, and new vaccines. Of note, the Department will place a major emphasis on developing improved vaccines for two serious and high-consequence bioterrorism threats, anthrax and smallpox.

The development of new or improved diagnostics, antibiotics, antivirals, and vaccines must go hand in hand with efforts to streamline the regulatory process. FDA will work closely with sponsors and manufacturers to ensure effective and timely reviews of investigational new products.

In conclusion, Mr. Chairman, I believe that the Department is making important progress to protect the health of this Nation from the threat of bioterrorism. Funding for this initiative this fiscal year totals $158 million. The President's request for fiscal year 2000 includes $230 million to continue to expand and strengthen the activities begun this year.

PREPARED STATEMENT

The medical and the public health communities clearly have the skill and the will needed for this task. We seek your help in ensuring that they also have the means. Thank you.

Senator SPECTER. Thank you very much, Dr. Hamburg.

[The statement follows:]

PREPARED STATEMENT OF MARGARET A. HAMBURG

Mr. Chairman and members of the Committee, thank you for the opportunity to testify today. I am accompanied by colleagues who have responsibility for implementing various parts of our anti-bioterrorism initiative: James Hughes, M.D., Director of the National Center for Infectious Diseases, Centers for Disease Control and Prevention; Robert Knouss, M.D., Director of the Office of Emergency Preparedness, Office of the Assistant Secretary for Public Health and Science; John Taylor, Esq., Senior Advisor for Regulatory Policy, Food and Drug Administration; and Anthony Fauci, M.D., Director of the National Institute for Allergy and Infectious Diseases, National Institutes of Health. The Department of Health and Human Services (DHHS) welcomes your interest in our efforts to develop effective counter-measures for possible uses of biological weapons against the civilian population.

I will outline for you the overall strategic approach that DHHS is pursuing in our anti-bioterrorism activities—emphasizing our efforts to strengthen the public health infrastructure for infectious disease surveillance related to potential bioterrorism agents and our efforts to enhance capabilities for medical and public health response should a bioterrorist attack occur. Following that, my colleagues and I will be pleased to respond to questions.

I begin by noting that bioterrorism presents a special set of challenges to our emergency preparedness systems, public health organizations, and consequence management capability. Unlike a bomb or discrete chemical exposure, a terrorist incident involving a biological agent may not be detected or even suspected until people begin to present with serious illness. This may occur at considerable and varying distance from the site of initial exposure, both in terms of onset of disease (incubation periods can vary) and geographic location (e.g., if exposure occurs in a transportation terminal, people can spread out widely before becoming ill).

With a bioterrorist event, there is also the possibility of concentric, spreading circles of communicable disease exposure, extending significantly the damage caused by the agent released. This kind of threat will also dramatically increase the level of public fear and potential for major civil disruption.

Increasing the urgency of the need for our nation to prepare for the potential threat of bioterrorism is the fact that the agents most likely to be used in this type of attack are pathogens not commonly experienced in this country or routinely dealt with by our medical system. This has a number of significant implications: (1) the population generally has little or no immunity to the pathogen and hence is more vulnerable (e.g., no longer vaccinated against smallpox); (2) medical providers generally are not familiar with the diagnosis and treatment of these disorders (which they may even fail to initially recognize); and (3) routine scientific research into the pathogenesis and treatment of certain of these disease conditions has been at very low levels compared to other agents of infection because they have not been perceived to be high priority or because they require levels of biological containment that are not available at most research centers. For these reasons, a sound strategy for addressing bioterrorism will be quite different from one that targets other types of terrorist acts.

The DHHS initiative features activities in five distinct but related areas: Deterrence of biological terrorism; Surveillance for unusual outbreaks of illness; Medical and public health response; Development of a national pharmaceutical stockpile; Research and development.

I will comment briefly on each.

Deterrence.—The Centers for Disease Control and Prevention (CDC) has the responsibility mandated by the Antiterrorism and Effective Death Penalty Act of 1996 to regulate the shipment of certain hazardous biological organisms and toxins (hereinafter called "select agents"). Organizations such as research universities, pharmaceutical manufacturers, and microbiological archives often have occasion, as part of their routine work, to send or receive samples of dangerous pathogens or toxins. DHHS regulations (42 CFR 72.6) require that all facilities sending or receiving shipments of select agents register with the CDC, maintain records of such transfers, and otherwise document their compliance. CDC's administration of the select agent rule is part of the Administration's multi-agency effort, led by the Department of Justice, to combat terrorism.

CDC also fosters safe design and secure operation of laboratories that handle select agents. This involves consultation with laboratory officials to help ensure that new, renovated, or proposed facilities meet standard guidelines for the infectious organisms that will be handled. Development of guidelines and training materials for use by laboratory personnel and provision of technical assistance to states as requested regarding their inspection programs for BSL 3 facilities also are part of CDC's responsibilities..

Surveillance.—Terrorist use of biological weapons against the civilian population is likely to be surreptitious. Absent an explosion, other immediate evidence of an attack, or notification of authorities by a perpetrator that an attack has been made (i.e., people have been exposed), the first responders will be health-care workers rather than fire or police personnel (as would be expected for a conventional emergency response scenario). The first indication that a silent attack has occurred probably will be an outbreak of some uncommon illness or an abrupt, significant increase in the incidence of commonly observed symptoms. How quickly the outbreak is detected, analyzed, understood, and addressed will determine the timeliness and effectiveness of the medical and public health response and hence the extent and severity of the impact upon the health and well-being of the affected community.

For example, most infectious agents have an incubation period measured in days or weeks. A silent release of a biological agent capable of producing a highly communicable disease, therefore, could afflict hundreds—if not thousands—of individuals over a wide geographic area during a period of several weeks before the need for a full medical and public health response could be identified and the response designed and mounted.

CDC is working to upgrade public health capability to counter bioterrorism through complementary, simultaneous improvements in the bioterrorism-related expertise, facilities, and procedures of state and local health departments and within the CDC itself. The emphasis areas are (a) preparedness planning by state and local health departments; (b) prompt reporting of cases of illness that might have been caused by terrorists; (c) epidemiological analysis of outbreaks to identify the source and mode of transmission; (d) laboratory identification and characterization of the agents causing the outbreaks; and (e) electronic communications among public health officials regarding occurrences of outbreaks and responses to them. CDC recently issued a competitive program announcement soliciting applications for cooperative-agreement awards whereby states and major metropolitan health departments can receive financial and technical assistance to effect desired improvements in one or more of the five emphasis areas. CDC will make these awards this summer.

Medical and Public Health Response.—Much of the initial burden and responsibility for providing an effective response by medical and public health professionals to a terrorist attack of any kind rests with the local governments, with support from state and federal agencies. Local public health systems almost inevitably will be

57-119 00 - 2

called on to provide protective and responsive measures for the affected populations, including:

—mass patient care—including the establishment of auxiliary, temporary treatment facilities or procedures for the movement of overflow patients to other geographic areas for care;

—in the case of a bioterrorist event, mass immunization or prophylactic drug treatment for groups known to be exposed, groups who may have been exposed, and populations not already exposed but at risk of exposure from secondary transmission and/or the environment;

—mass fatality management to provide respectful and safe disposition of the deceased, including animals; and

—decontamination of the environment.

Presidential Decision Directive 62 designates DHHS as the lead federal agency to plan and prepare for a national response to medical emergencies arising from the terrorist use of weapons of mass destruction. Within DHHS, this responsibility rests with the Office of Emergency Preparedness (OEP) within the Office of Public Health and Science.

OEP seeks to develop complementary medical response capabilities at local and national levels. It works closely with other agencies—especially the relevant components of the Department of Defense (DOD), the Department of Justice, the Department of Veterans Affairs, the Federal Emergency Management Agency, and others—with a view toward ensuring that plans for managing the medical consequences of terrorist acts are well integrated with other emergency response systems. To date, the anti-terrorism focus across the federal government has been on the prospect of nuclear or chemical attacks. Future preparedness efforts must focus on the prospect of bioterrorism as well.

In particular, OEP contracts with local governments for the creation of Metropolitan Medical Response Systems (MMRSs) and, within these agreements, is placing new emphasis on preparedness for mass patient care and other consequences of biological terrorism. Also, OEP is working to strengthen its four National Medical Response Teams and the National Disaster Medical System overall with respect to the bioterrorism threat so that they can augment local capabilities as needed in the event of an attack. To date, OEP has contracted with 27 municipalities to develop MMRSs. Another 8 MMRSs are to be initiated this year; plans to fund 12 more with redirected fiscal year (FY)1999 funds have been provided to the Subcommittee (bringing the total to 47); and the budget request for fiscal year 2000 includes $16.5 million for contracts with an additional 20 cities for MMRSs—bringing the total to 67. The long term goal is to establish MMRSs in all 120 metropolitan areas specified in The Response to Weapons of Mass Destruction Act of 1997.

National Pharmaceutical Stockpile.—A release of biological, and some chemical, weapons of mass destruction will require rapid access to quantities of pharmaceutical antidotes, antibiotics and/or vaccines that will not be readily available in the locations in which they would be needed unless special stockpiles are created. Because no one can anticipate exactly where a terrorist will strike and each local government does not have the resources to create sufficient stockpiles on its own, special stockpiles must be created and maintained as a national resource.

The initial focus will be on acquiring antibiotics useful in treating anthrax, plague, and tularemia; enhancing the utility of the existing supply of smallpox vaccine; and developing a cache of drugs and equipment for countering chemical attacks. Once research and development have yielded improved vaccines against anthrax and smallpox and new antiviral drugs effective against smallpox, they will be included in the stockpile.

CDC has responsibility for developing the stockpile. Fifty one (51) million dollars has been appropriated for this purpose this fiscal year, and a comparable sum is requested for fiscal year 2000.

Research and Development.—Capability to detect and counter bioterrorism depends to a substantial degree on the state of relevant medical science and technology. Without rapid techniques for accurate identification of pathogens and assessment of their antibiotic sensitivity, planning for the medical and public health response will be compromised significantly. Without efficacious prophylactic and therapeutic agents, even the best planned responses are likely to fail. The current base of science and technology is strong in some areas (e.g., certain classes of antibacterial drugs) and weak in others (e.g., rapid diagnostics, anti-viral drugs, and vaccines). Strong, sustained research and development in relevant scientific disciplines is the only proven way to remedy such deficiencies in knowledge and technology.

The National Institutes of Health (NIH) is reinvigorating its research related to the pathogenesis of—and host immune responses to—infectious organisms likely to be used in terrorist acts—e.g., the organisms that cause anthrax, tularemia, and plague. This research would be greatly facilitated by the acquisition of genome sequence information on these and related pathogens. The results of such genomics research—coupled with other pathological, immunological, biochemical, and microbiological information—are expected to facilitate pursuit of a variety of critical goals including the development of rapid diagnostic methods for the most likely biological weapons, the development of antiviral therapies for smallpox and Ebola virus, and the development of new vaccines for anthrax, cholera, and smallpox. NIH also will undertake an array of basic and targeted studies oriented toward development of new or improved methods to diagnose chemical exposures and determine their effects upon the nervous system.

Building upon the rapid advances of recent years in the molecular and cell biology of infectious organisms, the Department has requested $30 million in fiscal year 2000 specifically for developing improved vaccines for the highest priority bioterrorism threats: anthrax and smallpox.

Other DHHS agencies are engaged in relevant research and development as well. CDC, as part of the surveillance initiative I described earlier, is expanding its in-house Rapid Toxic Screen project to develop methods for measuring, within 48 hours, toxicants in human blood or urine samples. The goal over the next three years is to devise methods to identify and measure 150 different toxins and to achieve an in-house analytic capacity of 200 samples per day. As new methods come on line, CDC will disseminate them to state and local laboratories as appropriate for incorporation into their analytic repertoires. Also, FDA proposes to expand its research on detection and characterization of toxins that might be used by terrorists.

Looking more generally at the entire civilian medical response to chemical and biological terrorism, DHHS contracted in May, 1997 with the National Academy of Sciences' Institute of Medicine (IOM) to provide specific recommendations for priority research and development activities to improve that response. The IOM's report, delivered this past January, examines a wide range of research and development needs—including not only the medical response areas described above but also topics such as environmental detection of chemical or biological agents, personal protective clothing and equipment, and decontamination. My colleagues and I have found this to be an excellent and helpful study; and the Office of Science and Technology within the Executive Office of the President is using the IOM report as its framework for assessing and coordinating counter-terrorism-related research and development throughout the Executive Branch.

Expedited Regulatory Review.—The development of new or improved diagnostics, antibiotics, antivirals, and vaccines needed to combat bioterrorism must go hand in hand with efforts to streamline the regulatory process that new products must undergo successfully to be approved for marketing. FDA will work closely with sponsors and manufacturers to ensure effective and timely reviews of investigational new products. For example, NIH has created an Anthrax Vaccine Working Group, which brings together representatives of the NIH, FDA, and DOD to advance research and development relevant to developing a new anthrax vaccine. Also, FDA intends to accelerate the pace and increase the efficiency of its reviews by ensuring the availability of experts to guide sponsors through the regulatory process not only for new products but also for new uses of existing products.

In conclusion, Mr. Chairman, I believe that DHHS is off to a good start toward protecting this nation from those who would use biological weapons against the civilian population. Thanks to the leadership of President Clinton and the strong support of the Congress, the funding for the anti-bioterrorism initiative this fiscal year totals $158 million. Moreover, the President's request for fiscal year 2000 includes $230 million to continue, expand, and strengthen the activities begun this year. The medical and public health communities clearly have the skill and the will needed for this task. We seek your help in ensuring that they also have the means.

SUMMARY STATEMENT OF HENRY L. HINTON, JR.

Senator SPECTER. Our next witness is Mr. Henry Hinton, Assistant Comptroller General for National Security and International Affairs at the General Accounting Office. He is responsible for GAO's work at the Department of Defense and State, as well as U.S. intelligence and foreign aid services.

Senator Rockefeller and I, last June 17, requested GAO to do a study of the anthrax immunity program. We thank you for the results. Shortly thereafter, on July 1, Senator Rockefeller and I again requested a GAO study on the administration's threat assessment of biological and chemical agents.

Mr. Hinton, thank you for joining us. Again, both of those reports are available, and we look forward to your testimony.

Mr. HINTON. Thank you, Mr. Chairman. I am pleased to be here to discuss our ongoing work and preliminary observations on the biological terrorist threat and some aspects of HHS' bioterrorism initiative.

My comments this morning, Mr. Chairman, will address four issues: First, the intelligence agencies' judgments about the threat; second, the importance and benefit of threat and risk assessments; third, some preliminary observations about our ongoing work on the science behind the biological and chemical terrorist threat; and finally, I will provide some of our observations on the public health initiatives that deal with the new national pharmaceutical stockpile.

Let me turn to the threat. The U.S. intelligence community continuously assesses both the foreign origin and the domestic terrorist threat to the United States and notes that overall conventional explosives and firearms continue to be the weapons of choice for terrorists. Terrorists are less likely to use biological and chemical weapons than conventional explosives, at least partly because they are difficult to weaponize and the results are unpredictable.

However, some groups and individuals of concern are showing interest in biological and chemical agents. The possibility that terrorists may use biological and chemical materials may increase over the next decade, according to intelligence agencies.

We have previously reported on the value of using sound threat and risk assessments performed by a multidisciplinary team of experts for focusing programs and investments to combat terrorism. Without such assessments using sound inputs and a team of experts, there is little or no assurance that the programs and spending are focused in the right areas and in the right amounts.

As you mentioned, Mr. Chairman, we are looking into the scientific and practical feasibility of a terrorist or terrorist group improvising a biological weapon or device outside a state-run laboratory and program, successfully and effectively disseminating biological agents, and causing mass casualties. Much of the information we have obtained is classified and in the early stages of our evaluation.

Overall, our work to date suggests that, for the most part, there are serious challenges at various stages of the process for a terrorist group or individual to successfully cause mass casualties with biological agents. For example, a terrorist group or individual generally would need a relatively high degree of sophistication to successfully and effectively process, improvise a device or weapon, and disseminate biological agents to cause mass casualties.

HHS has not performed a formal, sound threat and risk assessment with a multidisciplinary team of experts to derive, prioritize, or rank in accordance with the most likely threats the Nation would face the specific items it plans to have in its pharmaceutical stockpile. Also, we note that several of the items HHS plans to procure seem to be geared toward the worst possible consequences from a public health perspective and do not match the intelligence agencies' judgments on the more likely biological and chemical agents a terrorist group or individual might use as explained to us.

PREPARED STATEMENT

Last, it is unclear from HHS' 1999 operating plan whether and to what extent the Department has fully considered the long-term costs, benefits, and return on investment of creating and sustaining the production and inventory infrastructure for such an initiative.

Mr. Chairman, that concludes my opening statement and I stand ready to address your questions.

Senator SPECTER. Thank you very much, Mr. Hinton.

[The statement follows:]

PREPARED STATEMENT OF HENRY L. HINTON, JR.

Mr. Chairman and Members of the Committee and Subcommittee: I am pleased to be here to discuss our ongoing work and preliminary observations on the biological terrorist threat and some aspects of the Department of Health and Human Services' (HHS) bioterrorism initiative. As you know, our ongoing work was requested by you in your capacity as the Chairman and Senator Rockefeller as Ranking Minority Member of the Senate Veterans Affairs Committee; Congressman Shays as the Chairman of the House Government Reform Committee, Subcommittee on National Security, Veterans Affairs, and International Relations; and Congressman Skelton as Ranking Minority Member of the House Armed Services Committee. Over the past 3 years, we have studied and reported on a number of issues concerning federal agencies' programs and activities to combat terrorism. A list of related GAO reports and testimonies is in appendix I.

It is frightening to think that a lone terrorist or terrorist group might be able to improvise a biological weapon or use other means to spread anthrax, smallpox, or other biological agents to cause mass casualties and overwhelm the health care system in the United States. There is no question that it would be unconscionable not to prepare to respond to, if not be able to prevent, such an incident. But some very important questions should be asked and answered as an integral part of any federal decision to invest in medical countermeasures or preparedness initiatives. This is one of those few areas in which national security and public health issues clearly intersect. It is also an area in which many disciplines of expertise must come together to perform the challenging tasks of assessing an emerging threat and focusing our investments on the most appropriate countermeasures and preparedness efforts.

My testimony will address four issues. First, I will briefly discuss intelligence agencies' judgments about the threat of terrorism. Second, I will highlight the importance and benefits of threat and risk assessments to provide a sound basis for targeting the nation's investments in combating terrorism—a widely recognized sound business practice we have discussed in our reports and testimonies.[1] Third, I will share some preliminary observations from our ongoing work on the science behind the biological and chemical terrorist threat, with some focus on biological agents. Finally, I will provide some of our overall observations on public health initiatives that deal with a new national pharmaceutical stockpile and the basis for selecting items to research, produce, procure, and stockpile for civilian defense against terrorism.

SUMMARY

The U.S. intelligence community continuously assesses both the foreign-origin and the domestic terrorist threat to the United States and notes that, overall, conventional explosives and firearms continue to be the weapons of choice for terrorists. Terrorists are less likely to use biological and chemical weapons than conventional explosives, at least partly because they are difficult to weaponize and the results

[1] See Combating Terrorism: Spending on Governmentwide Programs Requires Better Management and Coordination (GAO/NSIAD–98–39, Dec. 1, 1997); Combating Terrorism: Threat and Risk Assessments Can Help Prioritize and Target Program Investments (GAO/NSIAD–98–74, Apr. 9, 1998); and Combating Terrorism: Observations on Federal Spending to Combat Terrorism (GAO/T–NSIAD/GGD–99–107, Mar. 11, 1999).

are unpredictable. However, some groups and individuals of concern are showing interest in biological and chemical agents. The possibility that terrorists may use biological and chemical materials may increase over the next decade, according to intelligence agencies. While biological and chemical terrorism is still an emerging threat, many agencies have initiated programs and activities—with Congress' support and funding—to combat and prepare for this threat.

We have previously reported on the value of a new, post-Cold War approach of using sound threat and risk assessments performed by a multidisciplinary team of experts for focusing programs and investments to combat terrorism. Without such assessments using sound inputs and a multidisciplinary team of experts, there is little or no assurance that programs and spending are focused in the right areas in the right amounts.

We are looking into the scientific and practical feasibility of a terrorist or terrorist group improvising a biological weapon or device outside a state-run laboratory and program, successfully and effectively disseminating biological agents, and causing mass casualties.[2] Much of the information we have obtained is sensitive, classified, and in the early stages of evaluation. Overall, our work to date suggests that, for the most part, there are serious challenges at various stages of the process for a terrorist group or individual to successfully cause mass casualties with an improvised biological or chemical weapon or device. More specifically, our preliminary observations are that

—a terrorist group or individual generally would need a relatively high degree of sophistication to successfully and effectively process, improvise a device or weapon, and disseminate biological agents to cause mass casualties;

—a weapon could be made with less sophistication, but it would not likely cause mass casualties;

—some biological agents are very difficult to obtain and others are difficult to produce; and

—effective dissemination of biological agents can be disrupted by environmental (e.g., pollution) and meteorological (e.g., sun, rain, mist, wind) conditions.

For its part of domestic preparedness initiatives for combating terrorism, HHS received about $160 million in fiscal year 1999. These funds are intended for a variety of related preparedness efforts, including research and development and a new national stockpile for pharmaceuticals, millions of doses of vaccines for smallpox and anthrax, antidotes for chemical agents, and other items. For fiscal year 2000, HHS has requested $230 million for public health initiatives for dealing with bioterrorism. Our preliminary observations follow:

—HHS has not yet performed a documented, formal, methodologically sound threat and risk assessment with a multidisciplinary team of experts to derive, prioritize, or rank—in accordance with the most likely threats the nation will face—the specific items it plans to have researched, developed, produced, and stockpiled.

—Several of the items HHS plans to procure seem to be geared toward the worst-possible consequences from a public health perspective and do not match intelligence agencies' judgments on the more likely biological and chemical agents a terrorist group or individual might use.

—It is unclear from the HHS fiscal year 1999 operating plan whether and to what extent the Department has fully considered the long-term costs, benefits, and return on investment of creating and sustaining the production and inventory infrastructure for such an initiative.

THE FOREIGN AND DOMESTIC TERRORISM THREAT IN THE UNITED STATES

The bombings of the World Trade Center in 1993 and the federal building in Oklahoma City, Oklahoma, in 1995, along with the use of a nerve agent in the Tokyo subway in 1995, have elevated concerns about terrorism in the United States—particularly terrorists' use of chemical and biological weapons. The U.S. intelligence community, which includes the Central Intelligence Agency, the Defense Intelligence Agency, the National Security Agency, the Federal Bureau of Investigation, and others, has continuously assessed the foreign-origin and domestic terrorist threats to the United States. According to intelligence agencies, conventional explosives and firearms continue to be the weapons of choice for terrorists. Terrorists are less likely to use chemical and biological weapons, at least partly because they are more difficult to weaponize and the results are unpredictable. However, some groups and individuals of concern are showing interest in chemical and biological weapons.

---

[2] We recognize that some agents are communicable and could be spread without a weapon or device.

According to the FBI, there were 4 confirmed incidents of terrorism in the United States in 1992, compared with 12 in 1993, zero in 1994, 1 in 1995, 3 in 1996, and 2 in 1997. These incidents involved the use of conventional weapons.

THREAT AND RISK ASSESSMENTS CAN HELP DEFINE REQUIREMENTS AND PRIORITIZE AND FOCUS PROGRAM INVESTMENTS

We have pointed out that sound threat and risk assessments can be used to define and prioritize requirements and properly focus programs and investments in combating terrorism. Soundly established requirements could help ensure that specific programs and initiatives and related expenditures are justified and targeted, given the threat and risk of validated terrorist attack scenarios as assessed by a multidisciplinary team of experts.

Several public and private sector organizations use formal, qualitative threat and risk assessments to manage risk and identify and prioritize their requirements and expenditures. For example, the Defense Threat Reduction Agency, the Department of Energy, and the Federal Aviation Administration use such assessments in their programs. In addition, the President's Commission on Critical Infrastructure Protection[3] recommended in its final report that threat and risk assessments be performed on the nation's critical infrastructures, such as telecommunications, electric power, and banking and finance systems. In fact, the Federal Emergency Management Agency strongly endorses the concept of risk assessment, as it is the key to predisaster hazard mitigation—the foundation of emergency management. Moreover, the Department of Energy has stated that domestic preparedness program equipment purchases should be delayed until a risk assessment is completed to ensure that appropriate equipment is obtained.

Threat and risk assessments are grounded in a new, post-Cold War approach to thinking about and dealing with security issues called risk management. Risk management is the deliberate process of understanding "risk"—the likelihood that a threat will harm an asset with some severity of consequences—and deciding on and implementing actions to reduce it. Risk management principles acknowledge that (1) while risk generally cannot be eliminated it can be reduced by enhancing protection from validated and credible threats and (2) although many threats are possible, some are more likely to occur than others. Threat and risk assessment is a deliberate, analytical approach that results in a prioritized list of risks (i.e., threat-asset-vulnerability combinations) that can be used to select countermeasures to create a certain level of protection or preparedness. Generally, because threats are dynamic and countermeasures may become outdated, it is sound practice to periodically reassess threat and risk.

The critical first step in a sound threat and risk assessment process is the threat analysis. The analysis should identify and evaluate each threat in terms of capability and intent to attack an asset, the likelihood of a successful attack, and its consequences. To perform a realistic threat assessment, a multidisciplinary team of experts would require valid foreign and domestic threat data from the intelligence community and law enforcement. The intelligence community's threat reporting on foreign-origin terrorism is often general and, without clarification, could be difficult to use. However, a multidisciplinary team of experts can use the best available intelligence information on foreign-origin and domestic threats to develop threat scenarios. The intelligence community could then compare the threat scenarios to its threat reporting and validate or adjust the scenarios with respect to their realism and likelihood of occurrence as appropriate.

OUR ONGOING WORK EXAMINING THE BIOLOGICAL AND CHEMICAL TERRORIST THREAT

On the basis of information we obtained and analyzed to date, a terrorist group or individual[4] would generally need a relatively high degree of sophistication to successfully and effectively process, improvise a device or weapon, and disseminate biological agents to cause mass casualties. John Lauder, Special Assistant to the Director of Central Intelligence for Nonproliferation, recently testified that "the preparation and effective use of biological weapons by both potentially hostile states and by non-state actors, including terrorists, is harder than some popular literature

[3] The Commission, a government-private sector body established in 1996, was to develop a national strategy to protect the nation's critical infrastructures from physical and computer-based threats.

[4] For the purposes of our work, we define terrorist(s) as a non-state actor not provided with a state-developed weapon.

seems to suggest."[5] Because we are in an open forum and our work is sensitive and preliminary in nature, my discussion will remain limited.

Our ongoing synthesis of information and technical data from recognized experts suggests that some exotic biological agents—such as smallpox—are difficult to obtain, and others—such as plague—are difficult to produce. Processing biological agents for effective dissemination to cause mass casualties requires specific, detailed knowledge and specialized equipment. Moreover, improvising a device or weapon that can effectively disseminate biological agents to cause mass casualties requires certain items that are not readily available. In addition, successful and effective dissemination of biological agents in the right form requires the proper environmental and meteorological conditions and appropriate energy sources.

That is not to say that casualties would not occur if less sophisticated means were used. For example, if an agent were dispersed in a less effective form using less effective equipment, some casualties might occur. However, under these circumstances, the potential incident would be less likely to cause mass casualties. What we have learned is that capability is a critical factor. Terrorists have to handle risk, overcome production difficulties, and effectively disseminate a biological agent to cause mass casualties. We continue to gather and evaluate data on these matters and plan to report to our requesters this summer.

PRELIMINARY OBSERVATIONS ON HHS' PUBLIC HEALTH INITIATIVES RELATED TO BIOTERRORISM

On June 8, 1998, the President forwarded to Congress a fiscal year 1999 budget amendment that included a proposal to (1) build—for the first time—a civilian stockpile of antidotes and vaccines to respond to a large-scale biological or chemical attack, (2) improve the public health surveillance system to detect biological or chemical agents rapidly and analyze resulting disease outbreaks, (3) provide specialized equipment and training to states and localities for responding to a biological or chemical incident, and (4) expand the National Institutes of Health's research into vaccines and therapies. The Omnibus Consolidated and Emergency Supplemental Appropriations Act (Public Law 105–277) included $51 million for the Centers for Disease Control and Prevention to begin developing a pharmaceutical and vaccine stockpile for civilian populations. The act also required that HHS submit an operating plan to the House and Senate Committees on Appropriations before obligating the funds. The fiscal year 2000 request for HHS' bioterrorism initiative is $230 million, including $52 million for the Centers for Disease Control and Prevention to continue procurement of a national stockpile.

Our preliminary work suggests that an ad hoc interagency health care group led by HHS has not yet performed a formal, documented threat and risk assessment to establish its list of biological and chemical terrorist threat agents against which it should stockpile. In fact, several of the items HHS plans to procure do not match intelligence agencies' judgments, as explained to us, on the most likely chemical and biological agents a terrorist group or individual might use. According to HHS officials, the group identified its list through a process of evolutionary consensus among federal and nonfederal health experts. Because HHS did not document its process or methodology, we have difficulty evaluating its soundness and comprehensiveness.

According to HHS officials, the interagency participants identified the list based on:

—agent characteristics such as transmissibility and stability,
—likely impact on population (i.e., can it cause mass casualties),
—availability of treatment, and
—whether the agent could be weaponized.

The group chose four biological agents for HHS' stockpiling initiatives—inhalation anthrax, pneumonic plague, smallpox, and tularemia (a bacteria)—because of their ability to affect large numbers of people (create mass casualties) and tax the medical system.

On the basis of our discussions with HHS officials, it is unclear to us whether and to what extent intelligence agencies' official written threat analyses were used in their process. According to the Joint Security Commission's 1994 report on Redefining Security, without documented threat information, countermeasures are often based on worst-case scenarios. Valid, current, and documented threat information is crucial to ensuring that countermeasures or programs are not based solely on worst-case scenarios and are therefore out of balance with the threat. While HHS

[5] Unclassified statement by Special Assistant to the Director of Central Intelligence for Nonproliferation on the Worldwide Biological Warfare Threat to the House Permanent Select Committee on Intelligence, March 3, 1999.

officials told us that they obtained information from various experts, including intelligence analysts, the ad hoc interagency group making the decisions comprised representatives only from the health and medical community. As a result, we have not seen any evidence that the group's process has incorporated the many disciplines of knowledge and expertise or divergent thinking that is warranted to establish sound requirements for such a complex and challenging threat and to focus on appropriate medical preparedness countermeasures.

As required in the appropriations act I mentioned earlier, HHS prepared an operating plan for its fiscal year 1999 bioterrorism initiative. The plan discusses numerous activities on which the fiscal year 1999 appropriations will be spent within four areas:

—deterrence of biological terrorism,
—surveillance for unusual outbreaks of illness,
—medical and public health responses, and
—research and development.

We have reviewed the unclassified version of the operating plan. On the basis of our review of the plan, it is unclear whether and to what extent HHS has fully considered the long-term costs, benefits, and return on investment of establishing the production and inventory infrastructure for such an initiative. The reason I raise the issue of return on investment is that, until a valid threat and risk assessment is performed, we question whether stockpiling for the items on the current HHS list is the best approach for investing in medical preparedness. In addition, the HHS plan does not clearly address issues surrounding (1) the long-term costs of maintaining an inventory of items with a shelf life or (2) the safety and efficacy of expedited regulatory review of new drugs and vaccines.

CONCLUSIONS

We see many challenges ahead for HHS as it continues to decide how to target its investments for this emerging threat. Many frightening possible scenarios can be generated. But the daunting task before the nation is to assess—to the best of its ability—the emerging threat with the best available knowledge and expertise across the many disciplines involved. The United States cannot fund all the possibilities that have dire consequences. By focusing investments on worst-case possibilities, the government may be missing the more likely threats the country will face. With the right threat and risk assessment process, participants, inputs, and methodology, the nation can have greater confidence that it is investing in the right items in the right amounts. Even within the lower end of the threat spectrum—where the biological and chemical terrorist threat currently lies—the threats can still be ranked and prioritized in terms of their likelihood and severity of consequences. A sound threat and risk assessment could provide a cohesive roadmap to justify and target spending for medical and other countermeasures to deal with a biological and/or chemical terrorist threat.

APPENDIX I—RELATED GAO PRODUCTS

Combating Terrorism: Observations on Federal Spending to Combat Terrorism (GAO/T–NSIAD/GGD–99–107, Mar. 11, 1999).

Combating Terrorism: FBI's Use of Federal Funds for Counterterrorism-Related Activities (FYs 1995–98) (GAO/GGD–99–7, Nov. 20, 1998).

Combating Terrorism: Opportunities to Improve Domestic Preparedness Program Focus and Efficiency (GAO/NSIAD–99–3, Nov. 12, 1998).

Combating Terrorism: Observations on the Nunn-Lugar-Domenici Domestic Preparedness Program (GAO/T–NSIAD–99–16, Oct. 2, 1998).

Combating Terrorism: Observations on Crosscutting Issues (GAO/T–NSIAD–98–164, Apr. 23, 1998).

Combating Terrorism: Threat and Risk Assessments Can Help Prioritize and Target Program Investments (GAO/NSIAD–98–74, Apr. 9, 1998).

Combating Terrorism: Spending on Governmentwide Programs Requires Better Management and Coordination (GAO/NSIAD–98–39, Dec. 1, 1997).

Combating Terrorism: Efforts to Protect U.S. Forces in Turkey and the Middle East (GAO/T–NSIAD–98–44, Oct. 28, 1997).

Combating Terrorism: Federal Agencies' Efforts to Implement National Policy and Strategy (GAO/NSIAD–97–254, Sept. 26, 1997).

Combating Terrorism: Status of DOD Efforts to Protect Its Forces Overseas (GAO/NSIAD–97–207, July 21, 1997).

Chemical Weapons Stockpile: Changes Needed in the Management Structure of Emergency Preparedness Program (GAO/NSIAD–97–91, June 11, 1997).

State Department: Efforts to Reduce Visa Fraud (GAO/T-NSIAD–97–167, May 20, 1997).

Aviation Security: FAA's Procurement of Explosives Detection Devices (GAO/RCED–97–111R, May 1, 1997).

Aviation Security: Commercially Available Advanced Explosives Detection Devices (GAO/RCED–97–119R, Apr. 24, 1997).

Terrorism and Drug Trafficking: Responsibilities for Developing Explosives and Narcotics Detection Technologies (GAO/NSIAD–97–95, Apr. 15, 1997).

Federal Law Enforcement: Investigative Authority and Personnel at 13 Agencies (GAO/GGD–96–154, Sept. 30, 1996).

Aviation Security: Urgent Issues Need to Be Addressed (GAO/T–RCED/NSIAD–96–151, Sept. 11, 1996).

Terrorism and Drug Trafficking: Technologies for Detecting Explosives and Narcotics (GAO/NSIAD/RCED–96–252, Sept. 4, 1996).

Aviation Security: Immediate Action Needed to Improve Security (GAO/T–RCED/NSIAD–96–237, Aug. 1, 1996).

Passports and Visas: Status of Efforts to Reduce Fraud (GAO/NSIAD–96–99, May 9, 1996).

Terrorism and Drug Trafficking: Threats and Roles of Explosives and Narcotics Detection Technology (GAO/NSIAD/RCED–96–76BR, Mar. 27, 1996).

Nuclear Nonproliferation: Status of U.S. Efforts to Improve Nuclear Material Controls in Newly Independent States (GAO/NSIAD/RCED–96–89, Mar. 8, 1996).

Aviation Security: Additional Actions Needed to Meet Domestic and International Challenges (GAO/RCED–94–38, Jan. 27, 1994).

Nuclear Security: Improving Correction of Security Deficiencies at DOE's Weapons Facilities (GAO/RCED–93–10, Nov. 16, 1992).

Nuclear Security: Weak Internal Controls Hamper Oversight of DOE's Security Program (GAO/RCED–92–146, June 29, 1992).

Electricity Supply: Efforts Underway to Improve Federal Electrical Disruption Preparedness (GAO/RCED–92–125, Apr. 20, 1992).

Economic Sanctions: Effectiveness as Tools of Foreign Policy (GAO/NSIAD–92–106, Feb. 19, 1992).

State Department: Management Weaknesses in the Security Construction Program (GAO/NSIAD–92–2, Nov. 29, 1991).

Chemical Weapons: Physical Security for the U.S. Chemical Stockpile (GAO/NSIAD–91–200, May 15, 1991).

State Department: Status of the Diplomatic Security Construction Program (GAO/NSIAD–91–143BR, Feb. 20, 1991).

GAO REPORTS

Senator SPECTER. We had some reports from GAO which had gone to the Committee on Health, Education, and Labor, Subcommittee on Public Health, not to us. We had a draft report also. Can you give me an approximation as to when GAO will conclude the studies which Senator Rockefeller and I have requested?

Mr. HINTON. Later this spring, Mr. Chairman, probably around the May-June timeframe.

Senator SPECTER. May-June timeframe?

Mr. HINTON. Yes, sir.

Senator SPECTER. Well, to the extent you can expedite that, we would very much appreciate it.

Mr. HINTON. Yes, sir, we certainly will try to.

Senator SPECTER. Dr. Hamburg, there is very heavy classification, as reported to me, on the number of agents which constitute biological threats. This subject came up at an earlier hearing conducted by the Veterans Affairs Committee and we have been trying to move to declassify. In the interest of caution, I will not say how many biological threat agents there are, but there is an approved vaccine, as I am advised, for only one, that is anthrax, and there are activities with FDA—only two other of six potential vaccines have been licensed. The four remaining are investigational drugs.

There are a large number of biological threats for which the United States has no vaccines. So that when someone makes the comment, as Senator Rockefeller alluded to earlier, that there is no threat, it is hard to understand how that could see print.

But what is the assessment of HHS as to our ability to start to develop vaccines for this very large number of biological threat agents?

Dr. HAMBURG. We see it as a very significant responsibility of the Department, of course working with our colleagues in other parts of the administration, the Department of Defense critically, and with private industry——

Senator SPECTER. Well, is the research being done principally by the National Institutes of Health or are there other agencies hard at work on the scientific research work?

Dr. HAMBURG. The NIH has played and will continue to play an important role, a growing role. DOD, particularly through USAMRIID, has a long history of research in this area and is actively engaged. Other parts of the administration are also involved in important aspects of our research agenda.

Senator SPECTER. Who is pulling it all together? That is one of the issues——

Dr. HAMBURG. Right.

Senator SPECTER [continuing]. Which Chairman Deutch and I are looking at on weapons of mass destruction. Who is pulling it all together?

Dr. HAMBURG. In terms of coordination of the overall research agenda, both identifying the goals and action plans for implementation, the Office of Science and Technology Policy in the White House is coordinating with the NSC that effort.

Senator SPECTER. That is the agency headed by Mr. Dick Clarke?

Dr. HAMBURG. Dick Clarke is coordinating the overall counterterrorism activities. The Office of Science and Technology Policy is part of the Executive Office of the President, but it is not part of the NSC. It is the science adviser to the President's Office. The NSC and OSTP are working in coordination, with OSTP having the designated lead on the biomedical research agenda.

Senator SPECTER. To what extent is private industry participating? I had a talk with Admiral Crowe and we are going to have on our second panel Mr. Robert C. Myers from the BioPort Corporation, which is developing a vaccine for anthrax or is manufacturing, producing a vaccine for anthrax. According to Admiral Crowe, private industry is not really pursuing this issue because of their liability problems.

It is a remarkable situation that for anthrax, a known threat where the Department of Defense has announced a plan to inoculate all the servicemen, that there is only one source of supply. Question: To what extent to your knowledge is the private sector really digging into this problem?

Dr. HAMBURG. Well, we are increasingly trying to engage the private sector. But as you point out, the liability issues are enormous.

Senator SPECTER. Increasingly trying to engage the private sector, but what is the private sector doing, if you know?

Dr. HAMBURG. Through contracts within DOD, there are relationships with private pharmaceutical companies to pursue development of a number of vaccine products.

Senator SPECTER. Dr. Hamburg, would you supply in writing to this Committee the following information: Specifically, what is being done in the private sector? What companies are working on what vaccines, and what is the stage of progress? What evidence do you have, if any, that other companies are not pursuing this subject because of concern for civil liability?

Specifically, what is the role of the White House Office of Science that you have referred to? What is happening with NIH? We can bring Dr. Varmus in separately, but since you are in charge of this we would appreciate it if you would take a look at that for us. What specifically is NIH doing, and could they do more?

The subcommittee on HHS took the lead in increasing their allocation by some $2 billion and this ought to be on the front burner. I would like to know precisely what they are doing in this line as you evaluate it, and also what the Department of Defense is doing.

[The information follows:]

RESPONSES TO QUESTIONS SUBMITTED TO DR. HAMBURG BY SENATOR SPECTER

Question. Specifically, what is being done in the private sector? What companies are working on what vaccines, and what is the stage of progress?

Answer. At the present time, Bioport is the only private sector company producing a licensed vaccine for a pathogen (anthrax) that could potentially be used as a biological weapon. DynPort, under contract to the Department of Defense, is in the early stages of developing a cell-culture based smallpox vaccine; this work has not yet entered the phase of human clinical trials. Several smaller companies have ongoing research and development programs on agents that also have the potential to be weaponized, but none of these efforts has progressed beyond the stage of pre-clinical testing in animal models.

Question. What evidence do you have, if any, that other companies are not pursuing this subject because of concern for civil liability?

Answer. While manufacturers and researchers are not required to share their rationale for the products they choose either to develop or not develop, there are incentives and deterrents that will influence their decisions. Incentives include the likelihood that a product will generate a profit, proceed successfully through the stages of development, and be accepted by consumers and health providers. The deterrents include, but may not be limited to, the cost of development, length of time for development, and liability issues.

Manufacturers may view significant off-label use as a potential liability. When a product is developed and licensed for one indication and a manufacturer is approached about an off-label use of the product, as in the case of vaccination to protect against a biological agent used in bioterrorism, the manufacturer may view the latter use as a potential liability since such use is not supported by adequate clinical data.

The liability issue is particularly significant with regard to the development and use of vaccines. The reason vaccines may be specially vulnerable relate to the issue of mandatory vaccination because, unlike other drugs prescribed for a disease from which an individual already suffers, vaccines are given to otherwise healthy individuals. The person being immunized may not be aware of the severity of the diseases that the vaccine is designed to prevent, leaving them more apprehensive of being vaccinated and less tolerant of adverse reactions associated with vaccination. This set of circumstances may result in the loss of public confidence in vaccines, thus increasing the likelihood of litigation against vaccine manufacturers.

Question. Specifically, what is the role of the White House Office of Science that you have referred to?

Answer. The Office of Science and Technology Policy (OSTP), a component of the Executive Office of the President, is coordinating an interagency effort to develop a unified, government-wide research and development agenda to improve counter-terrorism capabilities. Development of new and improved vaccines is a category of special interest. OSTP does not presume to direct the efforts of NIH or other agencies. Rather, OSTP facilitates communications among agencies on topics of common interest. With respect to vaccine development specifically, OSTP has helped DOD and DHHS identify complementary interests and expertise and avoid undue duplication of effort.

Question. What is happening with NIH? What specifically is NIH doing, and could they do more?

Answer. Several NIH Institutes are engaged in basic research on infectious agents—including several viruses, strains of bacteria and toxins—that have been identified as potential agents for use in bioterrorist attacks. In addition, the National Institute of Allergy and Infectious Diseases (NIAID) conducts and supports research on the development of therapeutics and vaccines for pre-and post-exposure prophylaxis and treatment, as well as diagnostics, for a wide variety of infectious agents, including several of those identified as potential biological weapons.

At the present time, NIH (primarily NIAID) maintains active research programs on anthrax, smallpox, plague, tularemia, Staphylococcal enterotoxins, viral encephalitides, and viral hemorrhagic fevers. The major emphasis, however, is on smallpox and anthrax. In this connection, NIAID recently formed working groups on smallpox and anthrax vaccines to coordinate the development and testing of potential second-generation vaccines. In addition to these efforts, NIH has participated in the following interagency activities related to bioterrorism:

NSC TASKING CONCERNING BIOTERRORISM PREPAREDNESS (1998)

Interagency collaboration in the development of a research agenda for the rapid development of diagnostic tools, treatments and vaccines for diseases caused by bioengineered agents.

Agencies: DHHS (OEP, NIH, CDC, and FDA) DoD, DoE, VA, and OSTP.

CONSULTATION ON ANTHRAX VACCINES (JANUARY 15,1998)

Agencies: DHHS (NIH [NIAID, NIDR, & NICHD] and FDA), DoD.

INTER-DEPARTMENTAL WORKING GROUP ON RESPONSE TO A WEAPON OF MASS DESTRUCTION (1998)

Use of unapproved pharmaceuticals for WMD response.

Agencies: DHHS (OEP, FDA, NIH, & CDC), DoD, and VA.

FIVE YEAR INTER-AGENCY COUNTER-TERRORISM AND TECHNOLOGY CRIME PLAN WORKING GROUP 4: CRISIS/CONSEQUENCE PLANNING AND MANAGEMENT

Agencies: DHHS, DOJ, FBI, ATE, FEMA, DOE, NRC, EPA, DOS, NSA, DOC, DOT, FAA, OSTP, CIA, Treasury, VA, DOI.

FIVE YEAR INTER-AGENCY COUNTER-TERRORISM AND TECHNOLOGY CRIME PLAN WORKING GROUP 5: PREVENTING AND RESPONDING TO TERRORISM INVOLVING NUCLEAR, BIOLOGICAL & CHEMICAL (NBC) WEAPONS

Agencies: DHHS, DOJ, FBI, ATE, FEMA, DOE, NRC, EPA, DOS, NSA, DOC, DOT, FAA, OSTP, CIA, Treasury, VA, DOI.

Recently, NIAID, participating in the National Security Council's tasking on the R&D aspects of bioterrorism preparedness, developed a research program to provide effective defense measures for the civilian population. The research plan addresses both short and long-term activities targeted at the design, development, evaluation, and approval of defensive measures against possible bioterrorist events. The essentials of this plan include design/testing of diagnostics, design/development/clinical evaluation of both therapies and vaccines, and basic research and infrastructure.

Question. What is DOD doing?

Other than the collaboration between NIH and the U.S. Army Medical Research Institute of Infectious Diseases (USAMRIID) on projects involving anthrax and smallpcx, DHHS is not in a position to speak to DOD's activities in vaccine development. The Committee may wish to contact DOD directly—e.g. Colonel Gerald Parker, Commander, USAMRIID, who is a highly knowledgeable about DOD activities related to preparation against bioterrorism threats.

BIOLOGICAL THREATS

Senator SPECTER. We will pursue with the Department of Defense the issue of declassification of these biological threats, because in my judgment the American people have a right to know what is happening here so that there be an appropriate response. I appreciate what the three of you have testified to, but I do not think there is anywhere near an appropriate response, and that is because the public is uninformed.

Senator Rockefeller.

Senator ROCKEFELLER. Mr. Chairman, just commenting on your statement, I am not sure that is the reason—that the public is uninformed. I think, of course, the public is uninformed, because actually GAO kind of brushes this thing aside, or appears to.

You have, if you count up all of the agencies at the State, local, and Federal levels, probably some 35 different agencies that need to coordinate, or parts, subsets, of agencies that need to coordinate.

I think, generally speaking, the public does not spend a lot of time thinking about bioterrorism. But I think it is precisely that that makes it more important for us to do so.

Dr. Hamburg, let me just put this to you. Dr. Henderson, who appears later, said that the production of smallpox vaccine stopped in 1980, with the exception of the one that the chairman referred to, and that that was kind of a perfect time for others, rogue nations, et cetera, to start producing it. Smallpox is considered a thing of the past. GAO dismisses it as being unimportant because some other agency has not declared it to be important.

It does not occur to me that anything is unimportant when it comes to either chemical or biological weapons, but everything needs to be inventoried, everything needs to be taken seriously.

We have, now, what—6 or 7 million doses of smallpox?

Dr. HAMBURG. Usable doses, yes.

Senator ROCKEFELLER. Yes, and it would take a year or two to get us to the point where we could begin to disseminate those, for production? Not the 6 or 7 million doses, but to make the necessary amounts?

Dr. HAMBURG. With the smallpox vaccine there are two important issues. One is the use of the existing smallpox vaccine supply, which as you know is limited to somewhere around 6 million usable doses.

But there is also the issue of, if we are going to expand capacity, should we not actually build on advances in biomedical science and produce an improved, better, safer, more efficacious vaccine? And that would require an investment of dollars and several years of research and development. But I think that there is widespread agreement within the scientific community that we need to invest in the creation of a second generation smallpox vaccine, that, while the risk of smallpox may be relatively low, the consequences of an exposure to smallpox in this Nation and around the world would be truly catastrophic.

We need to prepare against the threat of bioterrorism. To not do so I think would be irresponsible. It is a highly lethal disease. We know that smallpox has been used in various efforts to finally weaponize it as a bioweapon. There is no available treatment, limited supplies of vaccine. We have a population in this country and internationally that is essentially totally vulnerable, immunologically naive because we no longer vaccinate against smallpox and we do not have naturally occurring disease.

It is communicable person to person in a highly contagious way, requiring for appropriate infection control very strict respiratory isolation. A few cases would rapidly overwhelm our health care capacity to achieve that kind of respiratory isolation, virtually ensuring the ongoing spread.

So that I think, given the implications of what even a small smallpox exposure could mean in terms of taking a disease that is presently not a threat to the human population in terms of naturally occurring disease and making it again potentially endemic around the world, we believe that we have to prepare in a very serious and aggressive way.

Senator ROCKEFELLER. For everything, right?

Dr. HAMBURG. Clearly, with limited resources and competing priorities——

Senator ROCKEFELLER. But the theory is we would like to?

Dr. HAMBURG. Yes.

Senator ROCKEFELLER. It would just seem to me that it would be reasonably clear if we were not preparing to do something about smallpox, that others, then, who have ill will for us, would understand that and make the necessary accommodations.

Dr. HAMBURG. Given that in certain arenas we have limited medical tools available at the present time, we have to look to the future. Perhaps we will not always make the right choices, but we have to make certain research and development investments now so that we can be prepared to protect the health of the public.

Senator ROCKEFELLER. A quick question for you and a quick question for Dr. Kizer. This coordination thing is staggering. I mean, 50 percent of the public health agencies not having the computers to communicate with CDC is embarrassing, shocking, all the rest of it.

You know, we have all of these agencies and all of these problems and potentially a lot of time or potentially not a lot of time to prepare. Is there a best way of doing this—I mean, how does one gather so many agencies into a coordinated attack in something called government in the United States, philosophically as well as literally?

Dr. HAMBURG. It certainly is a challenge, and in this particular arena it requires bringing together communities of professionals that historically have not worked closely together. Public health and the medical community have not been colleagues with the intelligence community and the law enforcement community in an ongoing way.

I think the threat of bioterrorism really requires that we all work together and forge these new partnerships. The issues of collaboration and the effective utilization of limited resources to really address the most significant threats before us is one that we need to continue to work on, and I think that there are many, many examples of fragmentation of programs that would be greatly strengthened if we could coordinate.

I do think it is critically important to point out—and it will not surprise you, since I am representing public health and medicine for the Department of Health and Human Services—that the role of public health and medicine in addressing the threat of bioterrorism in particular has been dramatically underappreciated and underdeveloped because there has been the continuing framing of the so-called "chem-bio" threat as though it was one unified issue.

But I think that, as I hope I indicated in my testimony, the paradigm of bioterrorism is really very different, and public health and medicine will be the front-line responders, and the investments in increasing the robustness of the public health infrastructure for disease surveillance will not only greatly benefit our efforts to prepare against bioterrorism, but also against a range of naturally occurring infectious diseases that represent very serious concerns.

Senator ROCKEFELLER. Do public health agencies out there see this as a serious threat, or do they see it as sort of an unfunded mandate coming down on them from the Federal Government, something that they choose not to have to deal with because they do not have the resources to do it and we are not probably going to give them to them?

Dr. HAMBURG. I think there is growing concern and there is a real desire on the part of local public health departments and State and local governments to really begin the preparedness, planning, the investments in building the public health infrastructure, the medical consequence management capacity, and of course the concerns about having the medical tools we need to treat and prevent disease.

We have gotten enormous response to the requests for proposals that have gone out from the CDC just recently around the new money in the fiscal year 1999 budget. It should be noted that fiscal year 1999 is the first year that our Department did receive targeted moneys for bioterrorism. But there is an enormous desire to engage on this issue.

There is, of course, confusion in the profession about how real is this concern, how adequately can we in fact prepare. There are some who want to throw up their arms and say it is hopeless. There are others who want to roll up their sleeves and try to work on it.

But I think that we have made important strides forward, but we have a long way to go.

Senator ROCKEFELLER. Thank you, Dr. Hamburg, and please say hi to your father for me.

Dr. HAMBURG. Thank you.

Senator ROCKEFELLER. He happens to be one of the great men in America, so I had to say that.

Dr. Kizer, you referred to having had some experience in this, and I remember in—I think it was 1982 or 1983 or 1984—and I think actually it was in Oregon or someplace like that—that there was a cult group that contaminated water supplies, trying to win a local election or something. Is that what you were referring to?

Dr. KIZER. No, that sounds like something that would happen in West Virginia.

Dr. HAMBURG. It happened in Oregon, actually.

Dr. KIZER. Yes.

Senator ROCKEFELLER. You cannot say "ORR-e-gon." You have to say "ORR-e-gin." You get in trouble.

Dr. KIZER. As a former Oregonian, I will have to discuss your relocation plan at a later time.

Senator ROCKEFELLER. Actually, I meant to say Vancouver.

Dr. KIZER. The incident I referred to, in response to your question, involved the contamination of watermelons with the pesticide aldecarb in 1985. That involved many States in the West and ultimately over 1,100 people were poisoned and about a half a dozen deaths resulted from it.

Senator ROCKEFELLER. My time has run out, but I am actually determined to prevail on this point. Well, I will have to prevail later, but there was an incident in 1982 or 1983 when there was a cult group, whose name I cannot pronounce.

Dr. HAMBURG. Rajneeshi.

Senator ROCKEFELLER. Yes.

Dr. HAMBURG. In Oregon, and they used salmonella to contaminate——

Senator ROCKEFELLER. That was it.

Dr. HAMBURG [continuing]. A salad bar, in the hopes of——

Senator ROCKEFELLER. Also, they attempted to get into the water system, and, I thought, were successful.

Dr. HAMBURG. I am not aware of that.

Senator ROCKEFELLER. They did not, OK. But in any event, the point is, things can happen, even in this country. It does not have to be in the Tokyo subway.

In terms of emergency care, Dr. Kizer, you have a background in ER yourself. VA has very strict rules about which services they provide to veterans. Only now will VA be able to provide a uniform benefit package for all veterans, which may include emergency care. With your own background as an emergency medicine physician, how would you rate VA's capacity at this point in terms of emergency care?

Dr. KIZER. The VA has not historically been, nor is it at present, known for its expertise in acute trauma care. Most of our patients have medical problems, and I would rate its capability of dealing with complex medical problems as much higher than dealing with the typical knife and gun club trauma that you see in ER and some other places that I have worked in the past.

That—well, let me leave it at that. I was going to take it to where I think you might have been going, but I will defer to see whether you do.

Senator ROCKEFELLER. No, because my time has long since run out. But I appreciate your answer on that.

Thank you, Mr. Chairman.

Senator SPECTER. Thank you very much, Senator Rockefeller.

Picking up on a couple of comments from Senator Rockefeller—put the lights off. We are going to move on to the next panel. I just want to make a couple of concluding remarks.

The issue of public awareness in my judgment is always the dominant force to get governmental action. If the people are aroused and these stories appear in the media, there is more of a focus of attention. I have already instructed staff to prepare a letter for my signature to the Secretary of Defense, Bill Cohen, outlining at least my views about the public disclosure and the declassification.

Senator Rockefeller puts his finger on the key spot that our Commission on Weapons of Mass Destruction is dealing with, where should the central authority be. We are debating the Office of the

Vice President versus some special office in the national security administration. My own view is that it takes somebody like the Vice President, with that kind of clout, to bring all these agencies together. But of course he has to have the support of the President. But recently Presidents and Vice Presidents have been close together. Those are items we are going to be working on.

Dr. Kizer, just one word as to your situation. We thank you for your forceful advocacy for the veterans budget. There is just a glimmering, with OMB not approving your testimony today and the internal battles as to how much the veterans budget is going to be, that this is a fight which may have to be elevated to Congress. Members may have to take a hand here.

I do not want to get involved in your discussions within the executive branch to any extent that you do not want to mention them, but would you care to offer any explanation as to why OMB did not approve your statement and what is happening with respect to our efforts to have a significant increase in the veterans budget?

Dr. KIZER. I think those are two separate issues. I am not sure that I can comment on why OMB was reluctant to concur with the testimony that we had offered, and I would offer that opportunity to them to explain that, should that be appropriate. As far as——

Senator SPECTER. We are going to offer them the opportunity to explain it, too, and to include a copy of the statement they would not approve, so we can see why they would not approve it. Coincidentally, we will look at the statement.

[The information follows:]

U.S. SENATE,
COMMITTEE ON VETERANS' AFFAIRS,
Washington, DC, March 24, 1999.

Hon. JACOB J. LEW,
Director, Office of Management and Budget,
Executive Office Building, Washington, DC.

DEAR MR. LEW: On March 16, 1999, I chaired a joint Veterans' Affairs Committee-Labor HHS Subcommittee hearing to examine the roles and preparedness of the Department of Health and Human Services and the Department of Veterans Affairs (VA) to respond to a domestic chemical or biological weapons attack. I invited Undersecretary for Health, Dr. Kenneth W. Kizer, to testify on VA's behalf. Attached, you will find a copy of the letter of invitation sent to Dr. Kizer.

Much to my disappointment, Dr. Kizer showed up at the hearing with no prepared written testimony. OMB cleared a redacted version of Dr. Kizer's proposed testimony which would have provided insight on the greater role VA could bring to bear on the issue of a domestic WMD attack. I directed Dr. Kizer to provide me a copy of OMB's pass-back and found it stifling.

The issue of medical preparedness and response to a domestic attack by a chemical or biological agent is critical to the country's well-being. That is why I am at a loss as to why OMB is interfering in a Congressional inquiry into this matter.

I look forward to your response.

Sincerely,

ARLEN SPECTER.

EXECUTIVE OFFICE OF THE PRESIDENT,
OFFICE OF MANAGEMENT AND BUDGET,
Washington, DC, May 13, 1999.

Hon. ARLEN SPECTER,
Chairman, Committee on Veterans Affairs,
U.S. Senate, Washington, DC.

DEAR MR. CHAIRMAN: I am writing to respond to your letter of March 24, 1999, inquiring about the Office of Management and Budget (OMB) clearance of testimony prepared for Dr. Kenneth Kizer of the Department of Veterans Affairs (VA) for a March 16, 1999, joint hearing of the Labor-HHS Appropriations Subcommittee and the Committee on Veterans' Affairs on the roles of VA and the Department of Health and Human Services (HHS) in medical preparedness and response to a domestic attack by a chemical or biological agent.

I want to assure you that this issue is very important to the Administration. The Administration is concerned that adversaries will increasingly rely on unconventional strategies, such as attacks with weapons of mass destruction (WMD), to offset U.S. military superiority. Armed with these weapons, terrorists could inflict tremendous harm. While the Administration is working to deter and prevent such assaults, we must also be ready to manage the consequences of a WMD attack. Therefore, the counterterrorism activities of Federal agencies are the subject of a formal analysis each year in the preparation of the President's Budget. In addition, OMB participates in extensive interagency working groups throughout the year, led by the National Security Council, that address WMD preparedness issues. We believe it is very important that relevant testimony prepared for the Congress reflect this interagency process and present a coordinated Administration position.

OMB's clearance process is designed to ensure that the Congress receives coordinated testimony from the Executive Branch that accurately reflects Administration positions. The clearance process, as it currently operates, was essentially developed during the Roosevelt Presidency in the 1930s and established in its current form in 1939. The basic purpose of the central clearance process has remained the same during each of the subsequent Administrations—ensure that Presidential policies are reflected accurately before the Congress.

OMB received Dr. Kizer's draft testimony the day before the hearing and circulated it to concerned agencies to ensure that it was coordinated within the Administration and accurately reflected Presidential policies. After reviewing the edits proposed by the agencies, OMB asked VA to make changes to the testimony. After receiving the requested edits, VA staff informed OMB the evening before the hearing that the edits were unacceptable to Dr. Kizer and that he had decided to appear without a written statement. The staff at OMB acted properly, within the time available, to make sure the Congress would receive testimony accurately reflecting the President's policies. I hope this answers your question.

Sincerely,

JACOB J. LEW,
Director.

DRAFT PREPARED STATEMENT OF KENNETH W. KIZER

Mr. Chairman and Members of the Committees: I am pleased to appear before you today to discuss current and potential roles for the Department of Veterans Affairs (VA) in emergency management, in general, and incidents involving weapons of mass destruction (WMD), in particular.

At the outset, though, I want to commend you for calling this hearing to address the vitally important issue of WMD preparedness planning and for your leadership in recognizing the important role that VA could play in the domestic response to terrorist uses of WMD.

BACKGROUND

Before addressing issues related to WMD, let me first review VA's current emergency management roles and functions. VA was first formally assigned a federal emergency management role in 1982, when Public Law 97–174 tasked VA with ensuring the availability of health care for eligible veterans, military personnel, and the public during Department of Defense (DOD) contingencies, and natural, manmade, or technological emergencies. Within VA, the Veterans Health Administration (VHA) has responsibility for this emergency management function, and this is generally referred to as VHA's fourth mission. Within VHA, the Emergency Management Strategic Healthcare Group (EMSHG)—formerly the Emergency Management Preparedness Office (EMPO)—has lead responsibility for this mission.

EMSHG staff are charged with fulfilling several statutory or other mandated missions. With its headquarters located on the grounds of the VA Medical Center in Martinsburg, WV, a budget for fiscal year 1998 of about $7.6 million, and personnel located throughout the nation's major population and transportation centers, the EMSHG staff currently plan, coordinate, administer, and execute VHA's participation in six interrelated emergency management functions. These functions are:

—developing guidance to ensure continuity of operations for individual VA medical centers, and, more recently, VHA's Veterans Integrated Service Networks (VISNs).

—providing back-up medical care for DOD personnel in wartime;

—participating in the National Disaster Medical System (NDMS);
—participating in the Federal Response Plan for natural and man-made disasters;
—responding to natural and technological hazards; and
—participating in the Federal Government's continuity of government program.

Specific VHA mission responsibilities include the following:

VA Contingencies.—Through planning and coordination, EMSHG Area Emergency Management (AEM) personnel act to ensure continuity of operations at VA medical facilities during emergency conditions. AEM staff responsibilities also involve, prior to emergency conditions, developing, managing, and reviewing plans for disasters and evacuations, and coordinating mutual aid agreements for patient transfers

DOD Contingencies.—Since 1982, under Public Law 97–174, VA is the primary backup to the DOD healthcare system during times of war or other emergencies involving military personnel. In this regard, VA maintains plans to allocate up to 25 percent of its staffed operating beds for DOD personnel (casualties) within 72 hours of being alerted of DOD's need for such capacity. More specifically, 68 VA medical centers are designated as Primary Receiving Centers; these Primary Receiving Centers are tasked with being prepared to receive DOD patients and execute plans for early release and/or movement of VA patients to other VA Medical Centers designated as Secondary Support Centers. Other VA Medical Centers provide DOD installation support as needed. EMSHG maintains data on the availability of VA beds and would coordinate the receiving and transfer of DOD patients if needed.

National Disaster Medical System.—Under an interagency agreement signed in 1984, VA, DOD, the Department of Health and Human Services (HHS), and the Federal Emergency Management Agency (FEMA) are partners in administering and overseeing the NDMS, which is a joint effort between the federal and private sectors to provide backup to civilian health care in the event of disasters. NDMS maintains memoranda of agreement with civilian hospitals to provide staffed, acute care hospital beds for treating large numbers of patients as a result of major peacetime disasters, as well as additional backup during a military contingency. VA participates fully in NDMS. VHA representatives meet regularly with the other agency partners for planning and coordination. Overall policy is set by the NDMS Senior Policy Group, of which I am a member.

Federal Response Plan.—Under Public Law 93–288, enacted in 1992, VA assists state and local governments in responding to disasters, and increasingly events requiring security coordination. VA may be tasked to provide engineering services, mass care and sheltering, resources support, health and medical services, and urban search and rescue. Planning and coordination of VA participation is performed by EMSHG headquarters staff, with assistance from field staff as needed. Both headquarters and field staff are subject to deployment to disaster sites to assist FEMA and HHS in providing disaster relief. Clinical staff from other VHA elements may also be deployed to such sites. Support for these staff is prodded by EMSHG.

Since Hurricane Andrew in August 1992, VA has provided direct health and medical services in response to 19 mayor domestic disasters that have included earthquakes, hurricanes, floods, ice storms, and the bombing of the federal building in Oklahoma City, as well as special events involving the pre-positioning of medical assets for response to potential terrorist incidents. These events have included the 1996 Summer Olympics, the Economic Summit of the Eight, the 1996 Republican and Democratic Conventions, the recent Papal visit, and the President's State of the Union addresses. Over the past nine years, this support has involved the deployment of over 1,000 VA healthcare personnel, and the provision of substantial amounts of medical supplies and equipment, as well as use of VA facilities. Important to note is that with the abolition of the Public Health Service's direct care capability and with the devolution of military health care in recent years, VA has increasingly been looked to as the federal government's primary and principle response asset for disasters requiring a medical response.

Natural and Technological Hazards.—Under Executive Order 12657 (November 1988), VA responds to natural and technological hazards as a participant in the Federal Radiological Emergency Response Plan and is tasked, along with DOD, to respond to accidents at nuclear power stations and accidents involving radiological threats or injuries. EMSHG is responsible for planning and coordinating VA's participation to meet federal emergency management requirements and to provide coordination for federal deployments of VA medical assets.

Continuity of Government.—Under Executive Order 12656 (November 1988), EMSHG maintains VA-specific relocation sites in support of the continuity of government program during national emergencies. EMSHG staff assist in the maintenance of a relocation site in Martinsburg and necessary communication facilities for use by VA top management in the event these are needed to continue Federal government functions during a major national emergency.

VA INVOLVEMENT IN WEAPONS OF MASS DESTRUCTION PREPAREDNESS PLANNING

VA has been involved—both directly and indirectly—in weapons of mass destruction (WMD) planning since Presidential Decision Directive #39 was published in June 1995. This Directive established the crisis and consequence aspects of WMD incidents and included funds for HHS to purchase pharmaceuticals in support of four National Medical Response Teams (NMRTs) sponsored by the Public Health Service (PHS). VA currently stores and maintains four identical pharmaceutical caches for the NMRTs in Washington, DC; Winston-Salem, North Carolina; Denver, Colorado; and Long Beach, California.

Presidential decision directive #62

Since May 1998, VA has regularly participated in WMD planning in accordance with Presidential Decision Directive #62 (PDD–62), "Combating Terrorism." Included among the elements of this directive was a tasking to HHS, working with VA, to ensure adequate stockpiles of antidotes and other necessary pharmaceuticals nationwide and to train medical personnel in NDMS hospitals. To date, there have been no HHS requests under PDD–62 for additional pharmaceutical storage by VA at sites other than the existing four NRMTs.

With respect to training, PDD–62 tasks HHS, working with VA, to train medical personnel about WMD in NDMS hospitals. Currently, EMSHG has initiated the development of a WMD training program for selected NDMS hospitals this fiscal year. The training program is targeted for implementation in fiscal year 2000. Of note, in fiscal year 2000, HHS has the discretion to reimburse VA up to $1,000,000 for expenses incurred in conducting the WMD training program for hospitals participating in NDMS. The following table summarizes VA activities specifically related to WMD planning.

	Fiscal years—		
	1998	1999	2000 (projected)
Funds expended or committed, excluding salaries	$6,100	$41,320	[1] $1,056,842
Fulltime equivalent employees	1.15	2.29	2.29

[1] Includes the $1,000,000 HHS reimbursement.

There are significant considerations still to be addressed, however, regarding VA's potential to assist in WMD-related training efforts. For example, at this time I am not aware of any provisions for funding necessary decontamination and personal protective equipment at civilian hospitals, which are essential components of training for these incidents. Without such equipment, hospitals will be unable to accept "walk-in patients" who are contaminated, because to do so would imperil their staff and patients already one site. Clearly, "hands on" would greatly enhance readiness. Given that VA is developing the training program, addressing this matter would significantly improve our efforts in this regard.

VA ACTIVITIES AND ROLE IN PHARMACEUTICAL PROCUREMENT

In addition to developing the WMD training program for NDMS hospitals, EMSHG has assisted selected VA Medical Centers in WMD training initiatives. VA Area Emergency Managers also have been participating in state and local planning and preparation for WMD response in their role as community and mutual support coordinators. For example, at the request of HHS, VA recently provided training on treatment of WMD victims to selected hospitals in St. Louis prior to the Papal visit. VHA is currently involved in planning and coordination for the NATO–50 meeting in Washington, DC, where we will support the pharmaceutical cache pre-positioning and other medical support responsibilities in accordance with the 1992 Federal Response Plan. Important to note, however, is that training for all VA medical centers—just as with training for all NDMS hospitals—needs to be improved to assure baseline medical capabilities for WMD readiness.

With respect to pharmaceutical stockpiles, VA operates under an interagency agreement with HHS (attachment A). EMSHG works with VHA's Pharmacy Benefits Management Strategic Healthcare Group (PBM/SHG), Emergency Pharmacy Service to procure, inventory, store, manage, and distribute pharmaceuticals, nationally. The contents of the medical caches is determined by HHS; VA procures and then manages the caches.

PBM/SHG, working with the National Acquisition Center, purchases from select vendors the pharmaceutical and medical supplies, as identified and directed by HHS. VA has responsibility to replenish these items, as needed, based on usage and/

or potency longevity and to store the cache in a manner that ensures conformity to FDA and manufacturer storage regulations and recommendations. When deployment of a cache is required, PBM/SHG coordinates logistical matters with EMSHG. A copy of the general responsibilities of the Emergency Pharmacy Service responsibilities and of the storage and maintenance policy for the medical caches is found at attachment B.

WMD PREPAREDNESS PLANNING AND VA'S FUTURE ROLE

In your letter of invitation to appear at this joint hearing, you noted that you wished to explore VA potential for an expanded WMD role. I welcome the opportunity to discuss this with you, since I believe that VHA's potential utility and value in this regard has not received appropriate attention. Since becoming Under Secretary, I have undertaken various relevant efforts—to the extent possible within budgetary and management constraints—to improve VA's institutional capabilities and infrastructure in ways that would both enhance VA's contributions to the federal government's efforts to address WMD preparedness and enhance VA's ability to manage its current patient portfolio.

In this regard, I must emphasize that VA's medical care budget is dedicated to providing care for veterans. Consequently, as we noted in a previous report to Congress in response to Public Law 105–114, if VA is to assume any expanded role in WMD planning, it must be accompanied by additional resources to ensure that our first and primary mission of caring for veterans is not compromised.

Notwithstanding these resource issues, however, I think it is relevant to point out that VA can uniquely contribute in incremental ways to the federal effort to prepare for possible terrorist incidents involving WMD. In the area of physician training and expertise, for example, the federal government can leverage VA's major role in health professional training. As you know, each year more than half of all medical students and a third of postgraduate physicians in training (i.e., residents) receive some of their training at VA facilities. In addition, more than 40 other types of healthcare professionals receive training at VA medical centers every year. In this vein, during the past three years I have utilized this aspect of VA's mission to expand physician training slots that will increase the number of individuals with appropriate expertise in those disciplines necessary to address WMD preparedness, as well as to better address existing and projected needs in related areas involving veterans care.

Specifically, I directed that VA establish and support a new Medical Toxicology Special Fellowship (the expertise most pertinent to WMD incidents); this was initiated last year. In academic year (AY) 1997–98, VA supported 1 fellow; in AY 1998–99, there are 6 fellows. Additionally, Occupational and Environmental Medicine is the Accreditation Council of Graduate Medical Education-approved residency program providing the most relevant expertise, and in AY 1995–96 VA supported 4.25 positions in this specialty. In AY 1997–98, we supported 8.25 positions in Occupational and Environmental Medicine (AY 1996–97 was a transition year). In AY 1998–99, this number has grown to 14.37 positions.

Within VHA's primary medical care mission, I believe there lies significant untapped potential for us to support the federal government's capability to respond to WMD incidents. VA facilities are geographically dispersed in essentially all of the nation's major metropolitan areas. We have long-established relationships with 85 percent of the nation's medical schools and some 1,200 universities and colleges overall. Because we are part of local health care systems, we have established relationships with the local emergency medical services systems. We already participate in much of the state and local emergency planning activities, as well as coordinate training programs and exercises. In a word, there already exists a solid base upon which VA can build an expanded role specifically related to WMD.

As I noted earlier, VA manages four pharmaceutical caches for the NMRTs, but beyond this, VHA's infrastructure certainly lends itself to increasing such support in terms of procurement, storage, and management of pharmaceuticals at multiple sites for ready distribution wherever needed. In fact, in recognition of the key role a VA facility can play to address WMD preparedness planning, the Office of the Attending Physician of the U.S. Capitol signed a Memorandum of Understanding with VA in September 1998 to provide for the procurement, maintenance, and storage of a customized WMD cache, to be kept at the Washington VAMC (which also is one of the NMRT sites).

In this same vein, Public Law 104–201 directs and funds HHS to develop 125 local emergency medical systems, or Metropolitan Medical Strike Teams (MMSTs) in selected cities across the United States. Currently, 22 MMSTs exist in various stages of implementation. It has become apparent by the requests for support to

PBM/SHG and EMSHG that the strike teams do not have the infrastructure and training in place to facilitate the purchase, maintenance, and distribution of the WMD-related medical caches.

Moreover, I note that while VA's role and potential to support the national response to WMD has not been widely recognized within the federal government, I am advised that in cities where VA facilities have routinely been actively involved in local discussions, local HHS-led Disaster Medical Assistance Teams (DMATs) teams typically view VA as an important resource for training and logistical assistance.

Thus, I believe that VA's role could be enhanced, at minimum, to formally and proactively provide expertise and service to the expanding network of MMSTs and DMATs, rather than as a "resource of last resort." And beyond assistance with matters related to pharmaceuticals, these same VA assets can be drawn upon to improve the federal government's preparedness for WMD incidents in terms of training; medical and clinical capability; coordination with state and local agencies; provision of logistical and other support for other federal agencies; and research.

CONCLUSION

In closing, I would like to offer my personal observations on the current state-of-affairs for WMD readiness in the United States based on my experience as an emergency physician, medical toxicologist, public health official and as the former director of Emergency Medical Services for the State of California. In brief, this country is woefully unprepared for a terrorist incident involving WMD.

There is, however, much that can and should be done in the near term to better prepare us for such an inevitable event. As VA's Under Secretary for Health, I know the assets, expertise, and potential capabilities VA can offer to address WMD readiness; they are significant, though too often unrecognized and overlooked—or under appreciated. This is unfortunate, since just as there is no doubt that floods, earthquakes, hurricanes, tornadoes, and other natural disasters will strike the United States in the future, there is no question in my mind that a terrorist event involving WMD will occur in the United States. It is only a matter of when, where, and whether we will be prepared.

Dr. KIZER. As far as the budget, we continue to work with all involved parties to see that we have the resources to provide care to the many men and women who rely on the system for that care.

Senator SPECTER. Well, we will pursue that generalized answer on another occasion.

We thank you very much, Dr. Kizer, Dr. Hamburg, and Mr. Hinton. Mr. Hinton, you had your hand up a moment ago, just a finger. Did you want to make another statement?

Mr. HINTON. Yes, sir. I would like to just offer a couple comments in response to Senator Rockefeller's comments just a few minutes ago. This is an important area that we have been following over the last couple of years, and we have watched the funding go up, and the Federal response to this problem has been very, very significant. The message that I want to convey is our message is not that we should stop funding in this area.

What we see missing from the picture—and it kind of goes to the government's machinery, its response to the problem here—is that we have not seen threat and risk assessments that would kind of take you through each of the threats that we have, the likelihood and vulnerability around the threats, and a prioritization of what we need to be focusing the Federal machinery and the resources on, so we make sure we get to the highest priority parts of that threat list.

We have not seen that yet, and that has been an observation that we have had across the full spectrum of Federal programs and activities that have also been growing.

The second point that I would leave with you is that we provided a classified report, Mr. Chairman, to the Congress in December of this past year that raised some very significant command and control issues that I would want to make sure the Committee is aware of and make sure that you all have a copy of that before I depart.

Senator SPECTER. Thank you very much, Mr. Hinton, Dr. Hamburg, and Dr. Kizer.

We are going to move now to——

Senator ROCKEFELLER. Mr. Chairman, could I just make a comment on that?

So what you are saying, Mr. Hinton, is that you want all the ducks to be in order? You want No. 1 threat, No. 2 threat, No. 3 threat, and No. 13 threat, all of them put in strict uniform order, and then you want a nice blueprint for how this vast array of agencies is going to coordinate, before we start to take, to go more seriously into how we are going to protect ourselves?

Mr. HINTON. No, sir, not——

Senator ROCKEFELLER. I love the idea of waiting 2 years, if nothing happens. But do you contemplate such certainty on that?

Mr. HINTON. Senator, no, I do not see the detailed accounting in that way. What I am suggesting here is that we are bringing the complete apparatus, to include the intelligence communities, to bear in helping think through what the top priorities may be that we need to target that can cause the mass casualties we may be concerned about.

It is that type of analysis that we have not seen done, that would help us make sure we are not letting anything that we need to be dealing with fall through the cracks.

Senator SPECTER. Thank you very much, Mr. Hinton. We are going to have to move ahead to the next panel. We have to conclude this hearing in advance of 11 o'clock. So we thank you all very much.

NONDEPARTMENTAL WITNESSES

STATEMENT OF JOSHUA LEDERBERG, M.D., PRESIDENT-EMERITUS AND SACKLER FOUNDATION SCHOLAR, ROCKEFELLER UNIVERSITY, NEW YORK CITY

Senator SPECTER. We would now like to turn to Mr. Joshua Ledberg——

Senator ROCKEFELLER. Lederberg.

Senator SPECTER. Dr. Joshua Lederberg, Dr. Donald Henderson, and Dr. Robert Myers. Our first witness on the second panel, Dr. Lederberg, is President Emeritus and Sackler Foundation Scholar of the Rockefeller University of New York City. No wonder Senator Rockefeller was so particularized about pronunciation.

Senator ROCKEFELLER. Pronounce the name correctly. He is a Nobel Laureate.

Senator SPECTER. I quite agree with your preeminence, Dr. Lederberg. I would try to pronounce everybody's name correctly even if they are not of your status.

Educated at Columbia and Yale, pioneered the discovery of genetic recombination of bacteria. In 1958, Dr. Lederberg was awarded the Nobel Prize in Medicine at the age of 33.

Thank you for joining us. Director Deutch sends you special greetings. We look forward to your testimony.

Dr. LEDERBERG. Thank you, Chairman Specter, Senators. Let me say right off I applaud your remarks at the beginning and the very bipartisan spirit with which they are given. My own may be preaching to the choir, but there are a few points I would like to emphasize.

First of all, before there is an aspiring reporter who wants to win a Pulitzer Prize for an astounding revelation, let me put it on the record, what everybody knows. I do consult quite widely in the biotechnology and the pharmaceutical industry and for SAIC, all of which would like to play a role in biodefense. I have also been for 30 years activistically involved in trying to deal with the threats of emerging infections and bioweapons.

The dissolution of the Soviet Empire and the refinement of our own military technology have vastly altered the security environment of the United States. Our forces in place have obliged our potential adversaries to avoid confrontation on discernible battlefields and to seek asymmetric strategies to avoid defeat upon head-on collision with us.

Biological weapons, germs, deployed against symbolic targets, troop barracks, and real valued ones like ports of assembly and cities, are almost ideally suited for that aim—post-modernist warfare.

With respect to major bio attacks by state interests, our principal bulwark remains deterrence. The main challenge is the credibility of our resolve to respond and the cementing of a global coalition dedicated to the enforcement of BW disarmament. That requires more work to win the minds and hearts of our friends as well as to define and lay down the law for the culprits. These may be the most vital steps, but they are an arena probably beyond the scope of these hearings.

But deterrence is futile against unattributable groups or crazies. BW can be delivered by clandestine vehicles, man or truck or boat-mounted sprayers, and can use prepositioned stockpiles measured in pounds, not tons. A major attack might hide below our radar screen of epidemic awareness for hours, if not days. Who then could be the target for our revenge in a deterrence mode?

Undeviating reliance on deterrence also leaves us open to provocation and disinformation from culprits and third parties to elicit responses that might discredit our legitimacy or even to catalyze war between tensely positioned states. Consider the interference with the peace process in Israel on the part of Hamas and its use of terrorism.

Nevertheless, it is smaller groups who are most likely to resort to BW, even if they lack access to unlimited resources from a mobilized state sponsor. Such smaller groups, which I will define as operating on a budget of a million dollars or less, still have readily achievable, although limited, capabilities.

Not enough attention has been paid to soft kill of urban targets, even while there has been some hyperbole about the ease of a hard kill of a city. For example, no one can dispute the feasibility of a scenario which could achieve 1,000 to 10,000 mortal casualties in a metropolitan area, far short of destroying the entire city, as is sometimes advertised.

What has been underestimated grossly is the terror that this would induce while the seeds of infection were germinating and spreading. In the wake of such an attack, even if the ultimate casualties added up to "only" 1,000 or 10,000, 100,000 to a million would be at risk, even according to the most rational calculus, because who would know who was infected and who was not. This could be amplified further by rumor and panic soon after the earliest cases came to notice.

Hence, even such—and I put this in quotes—"limited" attacks demand preparations for managing situational awareness, diagnosis, prophylaxis, and treatment for up to a million people—a staggering number and an event that cannot possibly be managed without a great deal of prior care, attention, mobilization, exercises, prepositioned stockpiles. It will be very difficult to know exactly who has been exposed in time to intervene and many more people would have to be cared for, many more people would have to be treated, than will in the end succumb, than will in the end have actually been exposed.

To deal with these issues, the public health infrastructure is the most important component. But this has to be designed and exercised to coordinate with all other elements of emergency management—public information, law enforcement, and if need be on such a scale, support from forces that can be mobilized under military discipline. Besides the structural arrangements would be provision for materiel, diagnostics, antidotes, hospital support equipment, including improvised beds, shelter, isolation, and so forth.

A brutal calculus might infer that this is demanding an investment of thousands of dollars per life actually saved, and this has to be qualified further by the unlikelihood that such attacks will really eventuate, or we do not know the likelihood. My answer is that this scale of investment is within the bounds of other prophylactic programs, but also that the penalty of a successful attack goes beyond thc lives lost. It will be an example for others, inviting multiplied problems. It will discredit the sensitivity and competence of government. The secondary damage from chaos and panic may exceed the primary kill, including, may I say, the deterioration of the taxes that can be collected on uninhabitable properties.

The secondary gains from a preparedness program include a form of deterrence against culprits by offering less obviously naked targets. Hence their incentive to use this mode of attack would be greatly lessened to the extent that we are in fact prepared for it.

Senator SPECTER. Dr. Lederberg, could you summarize the balance of your statement, and your full statement will be made a part of the record.

Dr. LEDERBERG. Yes.

PREPARED STATEMENT

Navy Secretary Danzig has characterized bioweapons as weapons of mass disruption, and for this most likely level of attack that is the appropriate perspective and should guide our priorities in allocating the resources for the defense of our cities.

Thank you very much.

Senator SPECTER. Thank you very much, Dr. Lederberg.

[The statement follows:]

PREPARED STATEMENT OF JOSHUA LEDERBERG

This writing is adapted from my introduction and epilogue to "Biological Weapons: Limiting the Threat" MIT Press 1999 edited by Joshua Lederberg.

The transcendence of BW over medicine and public health, private criminal acts, terrorism, interstate warfare, and international law directed at the elimination of BW, makes this one of the most intricate topics of discourse, poses very difficult security problems, and opens some novel challenges in the ethical domain. (See Table 1.)

That same transcendence confounds efforts to organize governmental and intergovernmental measures of control: health authorities will need to negotiate with the military, with law enforcement, with environmental managers. And all will have to cope with how to enhance security without imposing intolerable stresses on personal liberties and on freedom of travel and of commerce.

The topic of biological warfare (BW) had last been covered systematically by JAMA, as part of a discourse on weapons of mass destruction, in August 1989. [1 2 3] We recall that 1989 marked the bicentennials of the American presidency and of the French Revolution. By year's end, perhaps not by pure coincidence, 1989 also marked the collapse of the Soviet Empire, and with that the end of the cold war. The Biological Weapons Convention (BWC) had been in place since 1972; nevertheless, compliance on the part of great states, notably Russia, with that convention was the centerpiece anxiety in 1989. United States national policy was likewise concentrated on the defense of our troops in tactical combat settings.

Medical interests, notably symbolized by the World Health Organization's pleas [4] had played a significant role in the diplomatic priority given to the BWC, and then to concern for its enforcement. Since 1989, the Persian Gulf War, the escaltion of terrorism, and a recrudescence of many infections have added new dimensions to concerns for the malicious incitement of disease. Iraq was proven to have developed and militarized a repertoire of BW agents, notably anthrax spores.[5] Terrorists achieved new levels of violence in New York, Oklahoma City, and Tokyo and operated on ever more incomprehensible and unpredictable rationales. Having deployed chemical weapons in Tokyo and dabbled in BW, terrorists would soon be attempting to deploy BW on an increasing scale. It is not difficult to find recipes for home-brew botulinum toxin on the World Wide Web; terrorists justify this with the proposition that every citizen should have the parity of power with government. Meanwhile, the growth of biotechnology has great promise for new modes of diagnosis and therapy, but if left unchecked, advances in biotechnology will allow for even more troublesome microbiological agents of destruction.

This volume then touches on a set of timely concerns that unite national security and public health, concerns that cry out for well-articulated convergence of the human community worldwide. Various articles in this issue touch on the historical,[6] diplomatic,[7 8] and legal background;[9] on modalities of diagnosis and management;[10 11] and on case studies of small-scale BW attacks that have already been perpetrated, though amateurish in design and ending with limited malefaction.[12 13]

While BW is widely regarded as the absolute perversion of medical science, the problematics of invoking humanitarian regulation of means of warfare are well understood.[14 15] Resort to warfare is tied to the use of any means necessary for the survival of the state, including organized violence. It is mainly the peacetime behavior of states that can be regulated by international law, and this has evolved toward greater coherence and impact in an interdependent global economy. Even in the thrall of violent combat, states will also be deterred when there is a firm international resove: Iraq did not, after all, use its massive stockpiles of anthrax in the Gulf War.[5]

The 20th Century has seen the exercise of massive violence on an immense scale, even without major resort to BW. What distinguishes BW is the understanding that its habitual practice would be ruinous to personal security and civil order perhaps more grievously than any other weapon likely to get in the hands of disgruntled individuals or rogue states. One sine qua non for the elimination of BW is its utter delegitimation; in the language of the Geneva Protocol of 1925, it must remain "justly condemned by the general opinion of the civilized world."[16]

As a matter of international law, any debate has already been settled by the wide adoption of the 1972 BWC: the abnegation of biological weapons is approaching the status of a norm of international behavior, going beyond a mere contract for mutual compliance. When an international consensus can be achieved and sustained, as happened after Iraq's invasion of Kuwait, severe sanctions can be imposed by the international community. The task is to build that moral consensus[17] and give it sustainability and priority over more transient aspects of perceived national interest, like commercial advantage or access to resources. We are happily less burdened by the choosing up of sides of the cold war and the strange bedfellows that process engendered. There is much to answer for in the nonchalance exhibited by most of the world when Iraq used chemical weapons in its wars against Iran and its own Kurd dissidents.[18]

Writing 50 years ago,[20] Vannevar Bush remarked, in puzzling why BW had not been deployed at the height of World War II: "Without a shadow of a doubt there is something in man's make-up that causes him to hesitate when at the point of bringing war to his enemy by poisoning him or his cattle and crops or spreading disease. Even Hitler drew back from this. Whether it is because of some old taboo ingrained into the fiber of the race. . . . The human race shrinks and raws back when the subject is broached. It always has, and it probably always will." Bush could not offer these as reliable reassurances; and he surely played a large role in instituting and maintaining what became the US offensive BW development program. That started during World War II, and escalated in the cold war competition with the Soviet Union until 1969, with President Nixon's unilateral abnegation. In due course that was followed by the successful negotiation of the BW Convention of 1972, and its coming into force internationally in 1975.

Scrupulous adherence to the BWC on our own side, coming to the bar with clean hands, is of course an absolute prerequisite to the moral platform of BW prohibition. There is no more powerful instrument for that credibility than self-inspection. In free societies, that responsibility will largely devolve on well informed scientific and medical professionals. That community also has deep-seated ties with compeers even in some authoritarian states, bonds that should be cultivated to develop common ground even against obstacles of parochial interest.

Unlike nuclear weapons, the capability for BW is unlikely to be reliably contained by any degree of legal prohibition and formal verification. The facilities required for producing and dispensing BW agents are modest, easily concealable, and almost indistinguishable from licit production of pharmaceuticals and vaccines. The same holds for the underlying technical knowledge, which is part and parcel of medical research and education. The potential for grave enhancement[21] of virulence and intractability of pathogens for BW use goes hand in hand with the advances of biotechnology for human life-enhancement. Verification still plays a role, as part of a lawful process of investigation and indictment of malefactors. But the key to consolidation of the law on BW is its rigorous enforcement, and this will require a consensus even among our friends and allies that has yet to be achieved—partly out of the expectationthat the U.S. will always bear the onus as enforcer of last resort. Moral conviction and discreet technical education about the implications of leaving BW unchecked then go hand in hand.

As for the smaller and more marginal states, we should anticipate some ambivalence about foregoing weapons that might mitigate the overwhelming military power of a super-state. To enlist their unreserved cooperation in denying the use of BW, we should be far more proactive in mobilizing our health technology to stamp out rampant infectious disease globally.[22] Tuberculosis remains the earth's prime killer, and malaria with hundreds of millions of infected people the greatest drain on human vitality. It is scandalous that these coexist with a technology that will soon have plotted the entire human genome. Lacking robust technical solutions to the malevolent use of BW, we have little to call upon besides this common moral ground to prevent attack.

If despite deterrence, law, and moral suasion, the means of attack cannot be forsworn, there remains the obligation to be prepared to blunt them. Physicians and local health services, along with police and firefighter first-responders are in the front lines to deal with health emergencies. This is the same apparatus needed to deal with natural disease outbreaks: recall Legionella, Influenza A-H5N1, and Escherichia coli O157:H7 of recent vintage. The local responders also need to be trained in exercises entailing support from the Public Health Service and, if need be, military personnel. While BW attacks may be widely dispersed, far more than trauma from explosives or chemicals they are amenable to medical intervention: provided diagnosis is timely and resources can be mobilized. In many cases, there may be little or no advance warning. Vigilance in understanding the fate of victims near the dose-epicenter might provide an alert for the much larger cohorts likely to receive smaller doses and exhibit longer incubation times—a window of opportunity for treatment.

Several artiles in this issue point to recent progress, and a long way still to go, in the coordination of resources among a host of U.S. governmental agencies: federal, state, and local.[11 19] Recent press reports also speak to a rising tide of attention by responsible officials.[23]

In view of the rapid dispersal of people via jet aircraft, and the globalization of commerce, including foodstuffs, that coordination needs to be extended to a global venue. This scarcely exists at all at the present time, although the WHO has energetic programs to deal with influenza and HIV, and could be the nucleus of more extensive disease surveillance. With the growing recognition that BW is a strategic weapon, directed most effectively at large urban populations, cooperative public health measures might well reach the agenda of our security alliances like NATO. Military force protection against BW (and CW) is fairly advanced—with the dissemination of vaccines, antidotes and masks: these weapons are not likely to confer great tactical advantage to the perpetrator. Civil populations, near actual and potential theaters of combat or clandestine attack—and that no longer excludes our homeland—deserve comparable protection, if only to reduce the temptations for the aggressors, and soften the dilemmas and collateral harm of retaliation.

REFERENCES

Those for JAMA 1997 should be recast to refer:

[1] Orient JM. Chemical and biological warfare: should defenses be researched and deployed? JAMA. 1989;262:644–648.

[2] Sidel VW. Weapons of mass destruction: the greatest threat to public health. JAMA. 1989;262:680–682.

[3] Huxsoll DL, Parrott CD, Patrick WC III. Medicine in defense against biological warfare. JAMA. 1989;262:677–679.

[4] WHO Group of Consultants. Health Aspects of Chemical and Biological Weapons. Geneva, Switzerland: World Health Organization; 1970.

[5] Zilinskas RA. Iraq's biological weapons: the past as future. JAMA. 1997;278:418–424.

[6] Christopher GW, Cieslak TJ, Pavlin JA, Eitzen EM Jr.Biological warfare: a historical perspective. JAMA. 1997;278:412–417.

[7] Kadlec RP, Zelicoff AP, Vrtis AM. Biological weapons control: prospects and implications for the future. JAMA. 1997;278:351–356.

[8] Danzig R, Berkowsky PB. Why should we be concerned about biological warfare? JAMA. 1997;278:431–432.

[9] Ferguson JR. Biological weapons and the US law. JAMA. 1997;278:357–360.

[10] Franz DR, Jahrling PB, Friedlander AM, et al. Clinical recognition and management of patients exposed to biological warfare agents. JAMA. 1997;278:399–411.

[11] Holloway HC, Norwood AE, Fullerton CS, Engel CC, Ursano RJ. The threat of biological weapons: prophylaxis and mitigation of psychological and social consequences. JAMA. 1997;278:425–427.
[12] Toeroek TJ, Tauxe RV, Wise RP et al. A large community outbreak of salmonellosis caused by intentional contamination of restaurant salad bars. JAMA. 1997;278:389–395.
[13] Kovalic SA, Kimura A, Simons SL, Slutsker L, Barth S, Haley CE. An outbreak of Shigella dysenteriae type 2 among laboratory workers due to intentional food contamination. JAMA. 1997;278:396–398.
[14] Bull H. The Control of the Arms Race. New York, NY: Praeger for the Institute for Strategic Studies; 1961.
[15] Lederberg J. The control of chemical and biological weapons. Stanford J Int Stud. 1972;7:22–44.
[16] Protocol for the Prohibition of the Use in War of Asphyxiating, Poisonous or Other Gases, and of Bacteriological Methods of Warfare, (Geneva Protocol), 1925. Available at: www.opcw.nl.
[17] Cole LJ. The Eleventh Plague: The Politics of Biological and Chemical Warfare. New York, NY: WH Freeman; 1996.
[18] Kadivar H, Adams SC. Treatment of chemical and biological warfare injuries: insights derived from the 1984 Iraqi attack on Majnoon Island. Mil Med. 1991;156:171–177.
[19] Tucker JB. National health and medical services response to incidents of biological and chemical terrorism. JAMA. 1997;278:362–368.
[20] Bush, V. Modern Arms and Free Men. pp. 142, 146. New York: Simon and Schuster. 1949.
[21] Pomeransev AP., Staritsin NA., Mockov YV., Marinin LI. Expression of cereolysine ab genes in Bacillus anthracis vaccine strain ensures protection against experimental hemolytic anthrax infection Vaccine. 1997;15:1846–1850.
[22] Lederberg J. Infection emergent (editorial for theme issue). JAMA 1996;275:243–245
[23] Miller J., Broad WJ. Clinton Expected to Back Plan to Deter Terrorist Attack New York Times, April 26, 1998 (p. 1).

TABLE 1.—GERMS AS ARMS: BASIC ISSUES

BW vs CW: living germs vs chemicals germs might self-amplify and spread germs are biological unstable: could mutate to higher virulence
Underlying science is unalterably dual use licit defensive exploration targetted against natural disease
Likewise production up to point of weaponization vaccines vs. BW agents?
Facilities moderate scale; few external signatures easily concealed or masked by licit programs
Weapons: potent, but unfamiliar and unreliable in military context
Tactical defense is easy: physical barriers (masks, suits)
Latent period up to 36 hours. Disease may be treatable Hence focus on civil health preparedness
Hardly understood until now, these are *strategic* weapons. . . . At same time, accessible to small powers . . . or groups Seen as answer to a Superpower self confident about the "revolution in military affairs".
Capabilities can scarcely be denied remedial and intelligence focus on intentions

---

FROM EPILOGUE BIOLOGICAL WEAPONS—LIMITING THE THREAT: MIT PRESS 1999

(By Joshua Lederberg)

As these works were being assembled, our policy perspectives were informed by new happenings and governmental reactions. Saddam Hussein renewed his harassment of the UNSCOM inspectors seeking closure on Iraq's programs in BW and other weapons of mass destruction. Final rupture ensued, and with it renewed attacks on relevant Iraqi facilities on the part of the U.S. That escalation might be a deterrent/warning, or it might provoke unrasoned responses, including the use of BW if the regime inferred it had nothing more to lose. The dilemma persists how to invoke punishment on deviant autocrats without injuring captive populations even more severely. Nor is this forum the place to look beyond violence to the causes of belligerency. At one level, we knew the danger that violence will beget violence. At another, the history of nations has shown how the most violent exemplars, like Nazi Germany and Imperial Japan could eventually be pacified (and become models of pacific constraint and economic success)—at terrible cost to themselves and others. Democracies who regard themselves as humane will be torn, and sometimes self-deterred by such considerations, probably more than by threats of forceful retaliation. Saddam may not know this well enough to refrain from launching terrorist reactions; and there is always the cloak of fringe zealots acting on their own initiative.

This is the story line for the vicious bomb attacks on U. S. embassies in Kenya and Tanzania on August 8, 1998, attributed to Usama bin Ladin. Bernard Lewis [1] has retrieved bin Ladin's formal declaration of war against the United States and its citizens from the Arabic press. His aim is the expulsion of U.S. interests from the holy Arabian peninsula. In the process, hundreds of native Africans were injured or killed: this may go even beyond casual disregard of uninvolved bystanders—it conveys the message that diplomatic relations of any country with the U.S. entail a lethal liability.

This atmosphere has not triggered acute defensive precautionary mobilization beyond routine travel advisories. However, past months have witnessed a growing concern expressed in public pronouncements and official actions. Secretary William Cohen's foreword [2] is also reflected in President Clinton's Annapolis speech (May 22, 1998): ". . . three new initiatives—the first broadly directed at combatting terrorism; the other two addressing two potential threats from terroriss and hostile nations, attacks on our computer networks and other critical systems upon which our society depends, and attacks using biological weapons.

. . . We will work to upgrade our public health systems for detection and warning, to aid our preparedness against terrorism, and to help us cope with infectious diseases that arise in nature. We will train and equip local authorities throughout the nation to deal with an emergency involving weapons of mass destruction, creating stockpiles of medicines and vaccines to protect our civilian population against the kind of biological agents our adversaries are most likely to obtain or develop. And we will pursue research and development to create the next generation of vaccines, medicines and diagnostic tools. The Human Genome Project will be very, very important in this regard. And again, it will aid us also in fighting infectious diseases.

. . . To make these three initiatives work we must have the concerted efforts of a whole range of federal agencies—from the Armed Forces to law enforcement to intelligence to public health. I am appointing a National Coordinator for Security, Infrastructure Protection, and Counterterrorism, to bring the full force of all our resources to bear swiftly and effectively."

These decisions are reflected in Presidential Decision Directive [PDD–62], and the appointment of Richard Clarke of the National Security Council as the coordinator. Inter-agency discussions with regard to allocation of responsibility and budget are continuing. Significant announcements include the assignment of backup responsibilities to the National Guard [Cohen [2]]. The U.S. Atlantic Command (ACOM) already bears operational responsibility for "Homeland Defense", a theme much discussed in recent months, and it may be given further tasks in this arena. Not least is planning for the security of our ports of embarkation, the logistic chokepoints for maritime buildup and supply of any U.S force projection overseas.

Then, the Department of Justice will take over the training of local emergency responders to function safely and effectively in contaminated environments. Acting on its own, and impelled by past experiences like the attack on the World Trade Towers in 1993, New York City has already mounted an extensive program that will be a model for others.[3] In addition the FBI will establish a National Domestic Preparedness Office—a canonical shopping window for enquiries and appeals from local officials otherwise perplexed where to turn for assistance from the complex federal establishment. These proposals go a long way to meeting the criteria set out in a thoughtful paper by three recent members of the Clinton administration, Ashton Carter, John Deutch and Philip Zelikow.[4] They remark, however ". . . one should not place faith in czars. Real power still resides in the executive departments that have people, equipment, money, and the capacity to get things done." These requirements have been elaborated in further detail by Richard Falkenrath and his colleagues.[5]

Efforts to engage the Congress have been partly successful, but predictably face some resistance as "budget-busting" when incremental funding is sought. While there is substantial verbal endorsement of the priority that should be assigned to domestic bio-defense as an element of national security, it still fares poorly in competition with the long established traditional military concerns, the end of the Cold War notwithstanding.

The R&D requirements for bio-defense are barely touched upon in the current volume. They range from the most far-reaching innovations that will be called upon to deal with exotic viral infections to banal items like inexpensive, citizen-adapted protective masks. Protocols for the management of infectious disease were not designed nor validated for mass casualty settings, where for example available antibiotics are in short supply and rational schemes for extending those supplies will be desprately called for. Nor have our FDA and other regulatory and ethical regimes been confronted with emergent crises where thousands or millions of lives may be at stake, awaiting resolution of bureaucratic contradictions. Some of these matters have been given initial study by the Institute of Medicine.[6]

The delegation of responsibility to public authorities, and if so which ones, should be deliberated during times of peace, and informed consent conferred or denied; this cannot be achieved in the midst of crisis.

Among the triumphs of medical science and international cooperation in this century has been the global eradication of smallpox. Once among the major killers of humankind, smallpox has been eliminated from circulation by concerted programs of vaccination. The last authenticated case of naturally spread disease occurred in 1977; and the WHO officially declared eradication in 1979. Since then, accepted doctrine and general practice has been the abandonment of routine vaccination: the scourge had been lifted, no further precautions were needed. In consequence we now have, globally, a whole generation of humans with no history of exposure either to smallpox virus or to the protective vaccine. This is unprecedented in human experience, though it may be likened to the condition of Western Hemisphere natives prior to the European exploration and conquest. With recent rumor and defectors' reports of unabated experimentation with smallpox as a weapon, in defiance of the BW treaty, anxieties about our consequent vulnerability have been heightened. Outbreaks have happened before, and they could probably be contained—but only if vaccine stocks (now all but depleted) are refreshed and prepositioned.[7] This would not be very expensive; equally valuable and an important complement would be anti-viral medication if that could be materialized with renewed R&D.

My personal concern about the blight of biological weaponry, and the subversion of medical technology to the intentional spread of plagues, gos back many years. In 1970, I had occasion to address the United Nations Committee on Disarmament in Geneva, focussed on arms control as an important remedial device.[8] The treaty has been in place since 1975; it is now deeply embedded in the law of nations. The issue now is its enforcement, which depends on the institutionalized acknowledgment of and respect for that law, which is not to hide the problematics of focussing on a weapon rather than what is being fought over. BW is a special weapon, with implications for civility of life that set it apart from many other kinds of violence. Most of the other arguments remain hardly altered, except for the burgeoning realization of what biotechnology could bring us, for good or for evil.

FOOTNOTES

[1] B. Lewis, "License to kill," Foreign Affairs, Vol.77, No. 6. (1998), pp. 14–19.

[2] W. Cohen, "Foreword,"—this volume.

[3] Richard A. Falkenrath, Robert D. Newman, and Bradley A. Thayer, "America's Achilles' Heel: nuclear, biological, and chemical terrorism and covert attack," BCSIA Studies in International Security (Cambridge, Mass.: The MIT Press, 1998).

[4] Judith Miller and William J. Broad, "New York girding for grim fear: deadly germ attack by terrorists." New York Times, (June 19, 1998).

[5] A. Carter, J. Deutch, and P. Zelikow, "Catastrophic Terrorism," Foreign Affairs, Vol. 77, No. 6. (1998), pp. 80–94.

[6] Institute of Medicine, "Improving civilian medical responses to chemical or biological terrorist incidents. Interim Report" National Research Council, Washington. 1998.

[7] Joel G. Breman and D. A. Henderson, "Poxvirus dilemmas—Monkeypox, Smallpox, and Biologic Terrorism," New England Journal of Medicine, Vol. 339 (1998), pp. 556–559.

[8] Joshua Lederberg, "Address to Conference of the Committee on Disarmament, August 5, 1970." Congressional Record, (Sept. 11, 1970), pp E–8123–8124.

STATEMENT OF DONALD A. HENDERSON, M.D., M.P.H., DIRECTOR, CENTER FOR CIVILIAN BIODEFENSE, THE JOHNS HOPKINS UNIVERSITY

Senator SPECTER. We now turn to Dr. Donald Henderson, Director of the Center for Civilian Biodefense at Johns Hopkins University. He previously served as the Associate Director of the White House Science Office and the Senior Science Adviser to the Secretary of Health and Human Services.

Welcome, Dr. Henderson, and we look forward to your testimony.

Dr. HENDERSON. Thank you very much, Mr. Chairman and members of the Committee. I am really very pleased that you have taken the interest you have in this subject, which I think is a very important one.

My involvement in this whole field began in the Science Office in 1990 and has continued actively ever since. Throughout this period we have seen an escalating concern about the threat of weapons of mass destruction. But until recently, in the many meetings that I have participated in discussions about the implications of that threat and its possible scenarios have been confined primarily to those in the military, diplomatic, law enforcement, intelligence, and arms reduction communities.

Only recently have the civilian medical and public health communities begun to be engaged in examining the principal challenges posed by the threat. This has been a serious oversight.

So far, indeed, virtually all Federal efforts in strategic planning and training have been directed toward crisis management following a chemical release or an explosion and utilizing first responders, the fire, police, and emergency rescue workers. These groups do perform an important function and progress has been made in meeting this threat.

But a bioterrorism event presents an entirely different scenario. Unlike an explosive or chemical event, the bioweapons release will be silent and almost certainly undetected. Not until days or weeks later will patients begin appearing in emergency rooms and physicians' offices with symptoms of a strange disease which none of those seeing the patients have ever seen before. There will be no sudden alarm calling for action within minutes to hours on the part of "first responders," and in fact the first responders will not be fire and law enforcement staff at all.

Special measures will be needed for patient diagnosis, for care, for hospitalization, for obtaining the laboratory confirmation, for providing vaccine and perhaps antibiotics to large proportions of the population. We will need trained epidemiologists to identify where and when infection occurred so as to identify how and by whom it might have been spread, and we will have public administrators who will be challenged to undertake emergency management of a problem in an environment where, as Dr. Lederberg has said, the potential for panic is high.

In brief, the personnel that are required, the skills they must have, and the strategies to be employed could hardly be more different than those required for a chemical or explosive event. The prevailing assumptions which persist even today, that chemical and biological threats are so similar that they can be readily handled by multipurpose chem-bio experts, I think is as patently ridiculous as to assume that the athletes playing for the Washington Redskins are interchangeable with the Baltimore Orioles.

I was pleased last year when the Committee took cognizance of these challenges and appropriated to HHS and to CDC funds for planning, for strengthening the infectious disease surveillance network, and for enhancing the capacity of Federal and State laboratories. It is an important but modest sum of money, considering the needs of a fragile public health infrastructure extending over 50 States and 120 cities, none of whom are really strong at this particular point in time.

I would suggest to you that an augmented full-time cadre of professionals at this level, the State and local level, would represent for biological weapons a counterpart to the National Guard Rapid Assessment and Initial Detection teams, which are really more for chemical weapons. But the augmented capabilities to deal with such a bioterrorist event would not be on a standby basis, to be employed only in the case of emergency. Rather, they would be there working day by day, with the opportunity to deal with new and emerging infections and antibiotic resistance.

I should note that a risk assessment process has been in progress for nearly a year to identify which agents of potentially thousands are the most critical and for which we need to be prepared. Two in particular stand out as being of exceptional concern, smallpox and anthrax. Both are associated with high case fatality rates when dispersed as an aerosol. For smallpox we are talking about 30 percent, for anthrax 80 percent.

It is the view of our Hopkins group that reserve stockpiles both of anthrax and smallpox vaccines should be produced and antibiotics should be made available to deal with anthrax and with other agents as appropriate.

PREPARED STATEMENT

To come back to the point that I think that biologists, especially those in medicine and public health, are as critical to confronting the problems of biological weapons as are physicists and dealing with nuclear threats. They are sorely needed now to participate in a complex planning process, which you have identified, to blend together a very diverse array of institutions and talents in a way we have never done before, both public and private, in some sort of coherent plan. I believe it can be done.

Thank you.

Senator SPECTER. Thank you very much, Dr. Henderson.

[The statement follows:]

PREPARED STATEMENT OF DONALD A. HENDERSON

Good morning, Mr. Chairman and members of the Subcommittee. I am Dr. D.A. Henderson, Professor of Epidemiology and International Health and Director of the Center for Civilian Biodefense Studies at Johns Hopkins University. My involvement in counter terrorism activities date back to 1990–93 when I served as a Science Adviser to President Bush. I sustained this interest during my tenure as Deputy Assistant Secretary and Senior Science Advisor in the Department of Health and Human Services. I continue to serve on a variety of consultant panels to the DOD and the intelligence community.

This Committee has explicitly enunciated its own concerns with respect to bioterrorism. I share those concerns and am grateful for the support provided in helping us at Johns Hopkins to establish an academic center intended to catalyze a civilian planning and development program to deal with this vexing problem.

As this Committee is aware, this decade has been marked by escalating concerns about the threat of weapons of mass destruction. At first, discussions about the implications of this threat and its possible scenarios were confined primarily to those in the military, diplomatic, law enforcement and intelligence communities and to those concerned with arms reduction issues. Only recently have the civilian medical and public health communities begun to be engaged in examining the practical challenges posed by this threat.

THE UNIQUE NATURE OF THE BIOLOGICAL THREAT

Of the weapons of mass destruction, the biological ones are the most greatly feared but the country is least well prepared to deal with them. So far, virtually all federal efforts in strategic planning and training have been directed toward crisis management after a chemical release or an explosion. Should such an event occur, plans appropriately call for so-called "first responders"—fire, police, and emergency rescue workers—to proceed to the scene. There, with the FBI assuming lead responsibility, they are expected to stabilize the situation, to deal with casualties, to decontaminate, and to collect evidence for identification of a perpetrator. This is an important function and progress has been made in training and in developing such as the Metropolitan Medical Response Systems. Success in this effort must be attributed, in part, to the fact that spills or releases of hazardous materials, explosions, fires and other civil emergencies are not uncommon events.

A bioterrorist event presents an entirely different scenario, one that is alien to civil authorities. Epidemics of serious diseases such as are anticipated are wholly unknown to American cities. Unlike an explosive or chemical event, the bioweapons release would be silent and almost certainly undetected. The aerosol cloud would be invisible, odorless and tasteless. It would behave much like a gas in penetrating interior areas. No one would know until days or weeks later that anyone had been infected. Then, patients would begin appearing in emergency rooms and physicians' offices with symptoms of a strange disease that few physicians had ever seen. There would be no sudden alarm calling for action within minutes to hours on the part of "first responders". In fact, the "first responders" would not be fire and law enforcement staff but public health and medical personnel.

RESPONSES FOLLOWING A BIOWEAPONS ATTACK

Special measures would be needed for patient diagnosis, care and hospitalization, for obtaining laboratory confirmation regarding the identity of microbes unknown to most laboratories, for providing vaccine and perhaps antibiotics to large portions of the population, and for identifying and possibly quarantining patients. Trained epidemiologists would be needed to identify where and when infection had occurred, so as to identify how and by whom it may have been spread. Public health administrators would be challenged to undertake emergency management of a problem alien to their experience, in a public environment where epidemics of pestilential disease are unknown, and in an environment where the potential for panic is high.

In brief, the personnel that are required, the skills that they must have and the strategies to be employed could hardly be more different. The prevailing assumptions that chemical and biological threats are generically so similar that they can be readily handled by multi-purpose "chembio" experts is as patently ridiculous as to assume that the athletes playing on the Washington Redskins football team would be interchangeable with the Baltimore Orioles baseball club.

First responders to a biological weapons incident are emergency room physicians and nurses, family physicians, infectious disease specialists, infection control practitioners, epidemiologists, hospital and public health administrators, and laboratory experts. Until very recently, none of these groups had been meaningfully involved in assessing risks, nor in planning for appropriate civilian responses, nor in training, nor in defining research and development needs.

THE NATIONAL INITIATIVE

I was extremely pleased last year when the Committee took cognizance of these practical challenges in coping with bioterrorism and appropriated to HHS for fiscal year 1999, the sum of $133 million of which $51 million is intended for an emergency stockpile of antibiotics and vaccines. Other funds were allocated to the CDC, primarily for planning purposes, for the strengthening of the infectious disease surveillance network and for enhancing the capacity of federal and state laboratories. This was an important but still modest sum of money, considering the needs of a fragile public health infrastructure extending over 50 states and at least 120 major cities but it is an important beginning.

The provision of funds to HHS is consonant with the belief that one of the most critical elements for coping with bioterrorism is to strengthen the public health and infectious disease infrastructure. An augmented full-time cadre of professionals at the state and local level would represent, for biological weapons, a counterpart to the National Guard Rapid Assessment and Initial Detection Teams for chemical weapons. However, the augmented capabilities to deal with a bioterrorist event would not be on a standby status to be employed only in the case of an emergency. Rather, they would simultaneously serve to strengthen efforts directed toward dealing with such as new and emerging infections and food-borne diseases.

The sum of $1 million was identified for the Johns Hopkins Civilian Biodefense Studies Center. With these funds, a working group comprised of experts from federal, state and local institutions as well as academia have undertaken comprehensive reviews to identify the organisms that pose the most serious threats and that warrant special training of personnel in their identification and control and, as necessary, the development of relevant stockpiles of vaccines and antibiotics.

Two organisms in particular were identified as being of exceptional concern—smallpox and anthrax. Both are associated with high case fatality rates when dispersed as an aerosol. For smallpox, it is 30 percent; for anthrax, above 80 percent. Both have other advantages in that they can be grown reasonably easily and in large quantities and are sturdy organisms that are resistant to destruction. They are thus especially suited to aerosol dissemination to reach large areas and numbers of people. The working group supported the view that reserve stockpiles both of anthrax and smallpox vaccines should be produced and that a stockpile of antibiotics should be created to deal with anthrax. The group believed that the possible development of second generation vaccines for both diseases should be explored as a matter of urgency. Documents setting forth the consensus views of these experts should be published in a major national publication within the next few months. Other documents detailing the threat and response for at least four other agents will follow.

Over the past year, personnel from the Hopkins Center have made more than 50 presentations, on the request of professional organizations and hospitals, to acquaint them with the realities of the threat of bioterrorism and to discuss initiatives that need to be taken. On 16–17 February, a National Symposium, sponsored by the Hopkins Center, HHS and 12 other sponsoring organizations, took place in Washington. It was the first of its kind directed to a public health and medical audience. The response was so great that registration had to be closed 10 days before the Symposium convened. At this time, I believe it is reasonable to conclude that there is a markedly heightened concern and desire on the part of the medical and public health community to take a far more active role in the nation's preparedness.

A LOOK TO THE FUTURE

Biologists, especially those in medicine and public health, are as critical to confronting the problems posed by biological weapons as are physicists in dealing with nuclear threats and chemists with chemical weapons. There is a need to expand the discussion regarding issues both at national and local levels, to recruit the interest and commitment of scientists in devising strategy, in undertaking needed research and in the complex planning process which is needed to blend together the very diverse array of institutions, both public and private in coherent local, state and federal plans.

Plans for dealing with large numbers of patients, including those who require isolation will have to be elaborated on a regional basis and plans developed for emergency care facilities, for decontamination procedures, for dispensing rapidly large quantities of vaccine and antibiotics, for rapid and secure communications, for informing the media in a timely manner, for provision of mental health services and for emergency mortuaries.

Developing the experts and expertise will require a major educational effort, given the variety of specialists that are needed and the now virtual absence of knowledgeable and experienced specialists. There is a need to train primary care physicians and emergency room personnel in early recognition of the most important disease threats. Infectious disease specialists and hospital epidemiologists must also become versed in case recognition and in steps to take if a suspicious case is detected. There is a need for trained laboratory directors and key staff in laboratories with designated responsibilities for lab diagnoses. More over, state and local health officers and epidemiologists require training in, among other things, detection, surveillance and management of epidemic disease. Such an effort will require the full participation of professional organizations as well as those in the public sector and in academia.

Last but not least, it will be important to recruit the help of the medical and public health community in longer term measures that may prevent acts of terrorism. This would include strengthening the provisions of the Biological Weapons Convention Treaty and expanding our intelligence capabilities so as to anticipate and perhaps interdict terrorists. The fostering of international cooperative research programs to encourage openness and dialogue as is now being done with Russian laboratories is also important.

The possible role of the medical community in educating peoples and policymakers everywhere about the dread realities of bioterrorism has also been proposed as a parallel effort to an earlier initiative that proved so effective in clarifying the disastrous consequences of a nuclear war.

(For further information see: "The Looming Threat of Bioterrorism" in Science, 26 February 1999, pages 1279–1283.)

STATEMENT OF ROBERT C. MYERS, M.D., CHIEF OPERATING OFFICER AND DIRECTOR, BIOPORT CORP.

Senator SPECTER. We turn now to Dr. Robert C. Myers, Chief Operating Officer of BioPort Corporation, successor to the one company which produces anthrax vaccine. Dr. Myers was in charge of the State of Michigan's Division of Biological Products.

We appreciate your being here, Dr. Myers, and look forward to your testimony.

Dr. MYERS. Thank you very much, Mr. Chairman and members of the Committees.

I am the Chief Operating Officer of BioPort Corporation, the only FDA-licensed manufacturer of anthrax vaccine. I commend you to devoting today's hearings to bioterrorism and America's response to this critical issue. It is an honor to share my experience with you, an extended experience in biodefense vaccine development and manufacture.

I bring a somewhat unique perspective on the use of vaccines to combat bioterrorism. Unlike my colleagues, you have probably never heard my name before. That is no surprise. We do our work quietly. We try not to make waves.

The Defense Department came to us in 1990 to accelerate the production of anthrax and botulism vaccines in support of Desert Shield. I have been in Lansing for 21 years and for most of that time we were being asked to address the emergency need of one vaccine or another. For the last 10, it has been making sure there is anthrax vaccine to protect American troops being sent in harm's way.

We have worked closely with the FDA to get a quality vaccine up and running, and we have worked closely with the Army to build and test a stockpile of vaccine to meet an important DOD force protection requirement. In short, when it comes to both development and manufacture of biodefense vaccines, I know what it is about because I have done it. Now BioPort is the only company to manufacture a vaccine placed in routine use to protect against biowarfare.

What you have heard today should keep you awake at night. Tomorrow's biowarfare threats are today's responsibilities. Those threats are more real than some would like to think. Anthrax, as Dr. Henderson has pointed out, almost uniformly fatal. Smallpox, the entire world is susceptible and it is highly contagious. There are several others in the second string.

Heightening its threat, anthrax is a low tech bioweapon. It is easy to get, it is easy to grow—the recent hoax of choice. No one should question the potential for hoax to become reality sooner rather than later. As you have pointed out, Mr. Chairman, America just is not ready for that yet.

However, we have the power and knowledge to apply resources to diminish these threats to our safety, both here and abroad. But there is much to be done: intelligence, deterrence, detection, coordination, decontamination, containment, treatment.

It is complex to even think about from the perspective of a vaccine manufacturer. We have had an FDA-licensed anthrax vaccine since 1970, and just because it is old does not mean it is not good. How old are tetanus and diphtheria vaccines? We can be ready with a vaccine for anthrax. We have a good one now. We just do not have enough.

But we are not ready with large amounts of smallpox vaccine, and I believe it will take longer than the 3 or so years that some people are suggesting it will take to get a stockpile in place. A new smallpox vaccine, a new anthrax vaccine, or any other defense vaccine for that matter, faces serious challenge to licensure.

There has been little interest up until now from major pharmaceutical companies. The demonstration of both safety and effectiveness poses challenging issues for these unique vaccines. Critical manufacturing infrastructure will have to be fully developed and FDA licensed. This does not happen overnight. The stockpile will have to be generated and, as importantly, both the manufacturing sites and the stockpile storage sites will have to be protected against attack.

There will be criticism. Certainly the safety of a vaccine must be assured. Our experience with anthrax vaccine provides a perfect example of a safe vaccine that is still being criticized. It baffles me that someone going to the Middle East would actually refuse protection from a disease that is always fatal. You get inhalational anthrax, you do not get better; you die.

Vaccines given by injection do not get any safer than the anthrax vaccine. The side effects—a sore arm, a slight fever—occur less frequently than they do with common childhood vaccinations.

Bioterrorism is serious and demands that America take action. There is an enormous amount of work to be done and that work comes at a cost, both in time and money. But the greater cost is in failing to act in an informed, responsible, and reasonable manner.

I would be remiss if I walked away from this table without saying what some may perceive as controversial and self-serving. You are faced with choices that require almost unlimited resources, but your resources are limited. I strongly recommend that you focus those limited resources carefully. As far as vaccines go, take the FDA-licensed anthrax vaccine, manufacture it in greater quantities, and get it on the shelf and stockpiled for the civilian population now. Then get moving on smallpox vaccine now.

PREPARED STATEMENT

We are nowhere where we should be in protecting against what we once thought was eradicated from the face of this Earth. Concentrate America's vaccine-making resources where they are most needed and where they will do the most good. Tomorrow's biowarfare threats are indeed today's responsibilities.

Thank you. I would be happy to answer any questions.

Senator SPECTER. Thank you very much, Dr. Myers.

[The statement follows:]

PREPARED STATEMENT OF ROBERT C. MYERS

My name is Robert C. Myers and I am the Chief Operating Officer of BioPort Corporation. While I am not at the center of the policy and scientific discussions on how we move forward to best protect America against the scourge of biowarfare agents that could be used by terrorists and rogue nations, I am well qualified to speak to the development and manufacture of vaccines to provide the most effective shield of protection against this horrible sword of deranged aggression. Let me briefly explain.

As you probably know, BioPort manufactures the only FDA-licensed anthrax vaccine in the world. We are also making and testing a vaccine, under an approved Investigational New Drug (IND) sponsored by the Surgeon General, to protect against five different types of botulism, one of the most potent toxins known to man, a toxin that causes a miserable death by muscle paralysis. It is probably next on the threat list behind anthrax and the rapidly emerging threat of smallpox. I have been leading the Lansing, Michigan, facility's efforts to make, test and provide these vaccines to the Department of Defense for the last decade or so. Being at the front line of providing protection against these two weapons of mass destruction, we have grappled with the challenges that must be overcome to develop, make available and stockpile vaccines for which there is no market in the natural world—vaccines whose only market is derived as a response to the world of religious and political zealots and aggressor nations. These challenges can be overcome. I sincerely believe that the fastest and best approach to overcoming them is by first applying what we have learned up to now and then building on this.

The explosion of technological advances in medicine in the last decade have been thrilling to observe, especially in the area of new approaches to specifically stimulate the body's immune system to protect against and even treat all sorts of disease and sicknesses. At the same time, it has become alarmingly apparent that the likely near term threat is not based on burgeoning new technology but, rather, the evil and calculated use of organisms already known to nature. Most likely is anthrax. I say this because the organism is readily available, it can be easily grown to hugely destructive proportions in small, easily concealed labs, and weaponization techniques are well known, albeit not so easy to accomplish. Much less likely is smallpox because it is not readily available and its preparation as a biological weapon requires much more sophistication than that for anthrax. Nevertheless, because smallpox is highly contagious and probably most of the world is now susceptible, it is a potential biowarfare agent of serious concern.

There exist similar challenges to the further development and manufacture of new vaccines for anthrax, smallpox and, for that matter, any other biodefense vaccine. The first challenge is the challenge of interest. In this decade there have been few, if any, major pharmaceutical companies interested in developing and manufacturing defense vaccines. Their lack of interest was clearly established by their absence at the bidding table for the most wide-sweeping defense vaccine contract ever awarded by the Department of Defense—the Joint Vaccine Acquisition Program (JVAP). This program, to develop and license up to eighteen vaccines against important biological warfare agents and expected to take ten or more years at a funding of perhaps half a billion dollars, was well publicized, yet no major pharmaceutical company applied. I don't blame them. Such an effort is immense, will probably take longer and cost more than expected, and profits will be limited to that allowable by federal government regulations. With no clear second, more profitable market, there is little financial motivation to participate. Additionally, cumbersome government regulations and serious concerns about the government's ability to protect proprietary information probably cause large firms to shy away. There also is the greater likelihood of being subjected to an intrusive inspection under the Biological Weapons Convention, a likelihood made greater by the fact that those who make defense vaccines usually have the organisms, technology and many of the tools necessary to make the toxic and infectious components of biowarfare weapons. Under this scenario, important company proprietary information could and probably would literally "walk out the door" with the inspectors' records.

Perhaps most importantly, working on these vaccines could put a company's public image at risk. Consider the potential consequences for a large company, if it were making the anthrax vaccine—a vaccine that has been shown to be among the safest vaccines in the world. The unfounded adverse publicity surrounding this vaccine is great and it makes no sense for these companies to risk their reputations, be forced to defend frivolous litigation, or even experience product boycotts as a result of participating in a program such as JVAP.

Another challenge is embedded in the technical base of research and early development of any biodefense vaccine. In 1996, I was part of a team of organizations, led by Battelle Memorial Institute, which came together to compete for the JVAP award. We researched potential biodefense vaccines, which were being developed at the time. We found the science in the research and development of new vaccines to be generally excellent, with the proof of concept (the demonstration that a vaccine can be made and that it does protect against a target organism) well established for several. However, we often found that some of these vaccines were not as far down the path of development as the scientists would have liked us to believe. This challenge—the need for further testing, etc.—should be easily addressed by focusing on the real goal: on-the-shelf stockpiles of FDA licensed products. In addition to good science, the process also demands that specialists in manufacturing development, clinical trials, quality assurance, engineering and regulatory compliance be involved now. Full integration of these specialists, from the initial stages of each vaccine development project, is essential.

Once a vaccine product is ready to be tested in humans, there are special challenges for biodefense products that have yet to be addressed. Human Subjects Review Boards—those bodies that review and approve clinical studies from perspectives that include medical ethics—will have to determine the ethical acceptability of human studies of new vaccines that are only of hypothetical benefit—i.e., the vaccine's real benefit lies in a hypothetical situation that has not yet occurred—most likely a terrorist act. The approval of clinical trials under these circumstances is not a trivial matter as the circumstances of exposure are fundamentally different for civilian men, women and children than for service personnel whom we know are at risk of exposure to these horrific biological agents in battlefield conditions. I do not yet see a clear path to overcoming this ethical challenge. I am not an expert in this area and will leave this issue to the medical ethicists among us.

There are also special challenges to demonstrating both the effectiveness and safety of biodefense vaccines. For most of the vaccines that need to be developed, there is insufficient natural disease anywhere on this planet to demonstrate directly in humans that a given biodefense vaccine protects against disease. To overcome this challenge, animal and perhaps cell culture models of protection will have to be developed. These "models of effectiveness" will have to identify and demonstrate that there is a measurable, immunological marker that correlates well with protection. This marker must also be measurable in humans and be confirmed in human clinical trials as a surrogate marker of protection in lieu of the impossible direct vaccination of human, followed by exposure to the disease-causing organism. These studies must be accepted as valid by the scientific and medical communities, the FDA and ultimately the general public if any new vaccine is to have a place in preventing one of the terrible diseases that could be unleashed by a terrorist action.

The adequate demonstration of safety in humans is also a special challenge for biodefense vaccines. For these vaccines, we may have to change our view of the way safety is established. Human studies may have to be larger and more highly conclusive. I say this because the ongoing safety surveillance that occurs with most vaccines will be absent. Ongoing safety is routinely demonstrated for most vaccines because most vaccines are used regularly. In all likelihood, biodefense vaccines will not be generally administered; they will be used only in response to a terrorist or rogue nation biowarfare attack. Absent the ongoing safety surveillance, the scope and magnitude of human clinical trials may need to be expanded to increase the certainty that any biodefense vaccine we might use will be safe if and when it has to be used.

Other aspects of demonstration of clinical safety are also challenging. Unlike the services, where the recipients of biodefense products can reasonably be described as an adult population, exposure to a terrorist attack would include all age groups—babies, children, adolescents, adults and the aged. I see as formidable the challenge of planning and completing the necessary studies to confirm safety conclusively in specific age groups.

Further, the current experience with BioPort's very safe anthrax vaccine and the evolving FDA policy changes for all vaccines require that special studies be conducted that have not been traditionally required for vaccines in the past. These studies include evaluation in accepted models of the potential for toxicity to the fetus, fertility in both men and women, and perhaps even carcinogenicity. I would be the first to tell you that these studies are of questionable need scientifically for these vaccines. Unlike drugs, present vaccines largely exert their effect by stimulating the human immune system to respond specifically in one or more ways to a disease-causing organism. It is this stimulation of the immune system that later protects the individual against disease, when exposed to the organism. Also, unlike drugs, this ongoing immune response is the only lasting trace of the vaccine. There is no goal to sustain a certain level of the vaccine's ingredients, as is usually required of drug and other medicines. This continued exposure to drugs to treat medical conditions—often for one's entire life—is the basic reason why such studies are entirely appropriate for drugs and much less so for many vaccines. For decades we have vaccinated babies, children, adults and the elderly against tetanus without over this extended period of time finding any suggestion of specific fetal toxicity, impairment of fertility or cancer. In fact, in much of the world today pregnant women are intentionally vaccinated against tetanus to prevent neonatal tetanus in their babies, a disease still of serious concern in less developed parts of the world. Few if any vaccines have been examined in the ways for which anthrax vaccine is being specifically criticized today. In fact, two of the most recently licensed vaccines—vaccines for Hepatitis A and Hepatitis B—have not been fully evaluated for their effect on pregnancy. These vaccines are each made by two large pharmaceutical companies (a total of four products). The prescribing information for these four products each contains the same statement.

> Pregnancy Category C: Animal reproduction studies have not been conducted with ______. It is also not known whether ______ can cause fetal harm when administered to a pregnant woman or can affect reproduction capacity. ______ should be given to a pregnant woman only if clearly needed.

Remember, these hepatitis vaccines are among the newest, high-tech recombinant vaccines available today. All four contain the same statement that accompanies the anthrax vaccine, although only anthrax has received serious criticism. In fact, there are at least 25 FDA-licensed vaccines that carry the same statement. The fact that a vaccine has not been studied for its affect on fertility, the fetus and cancer does not mean that there are effects. It simply means that it hasn't been studied. This point is often lost in the din of criticism of anthrax vaccine in the popular press and I expect that new biodefense vaccines could easily become targets of such criticism as well. Some of these issues on safety are just plain baffling to me. I have received many doses of anthrax vaccine through the years. During this time, my wife and I have been honored by the miraculous gift of life three times. And I now have a three-year old grandson as well. If the anthrax vaccine were available today, I would have absolutely no reservation in administering it to my wife, grandson and three daughters, including my eldest who is of childbearing age.

Security will also be a challenge. As we improve our preparedness to combat bioterrorism, our preparation measures themselves will become targets of terrorist attack. Suppose we have a smallpox vaccine stockpile and a manufacturing capability. Suppose a terrorist group has smallpox as a weapon. While a successful terrorist action becomes more difficult, it is still possible and maybe simple. A bomb to the stockpile, a bomb to the manufacturing facility and a release of smallpox to a major city. A deadly one, two three combination. Funding for adequate security must be included in this program if the threat is to be optimally minimized. Included in these security measures and to prevent against natural disaster, there should be two or more geographical separate manufacturing facilities and two or more facilities for storage of the manufactured vaccine.

We collectively, all of us here today have a tremendous opportunity, responsibility and obligation to minimize exposure and risk to our citizens. Bioterrorism is a serious situation that demands that America take the offensive. There is an enormous amount of work to be done, and that work comes at a cost—both in time and in money—but the greater cost is in failing to act in an informed, responsible and reasonable manner.

I was invited here today to give my perspective and I would be remiss if I walked away from this table without saying what some may perceive as controversial and self-serving. You are faced with choices that require unlimited resources but your resources ARE limited. I strongly recommend that you focus those limited resources on those areas of bioterrorism in which we are most vulnerable. Specifically, take the FDA-licensed anthrax vaccine, manufacture it in greater quantities and get it on the shelf and available to the civilian population now. Then, get moving on smallpox. Now. We are nowhere near where we should be in fighting what we once thought was eradicated from the face of the earth. Concentrate America's resources where they are most needed and where they will do the most good.

Tomorrow's biowarfare threats are indeed today's responsibilities. Thank you.

COORDINATE EXISTING PROGRAMS

Senator SPECTER. Let me start with a question to you, Dr. Lederberg. As I say, we are running very close on time. We are going to have to conclude before 11. You outline an overwhelming problem which we all recognize, and great details as to what needs to be done. What is your suggestion as to how we structure the Federal Government to put somebody in charge who would have the clout to bring all of these disparate groups together and to provide the articulation for the public to get the kind of funding?

Dr. LEDERBERG. Senator, I have been wrestling with that problem at the philosophical level that you were engaging in earlier for at least 5 years and trying to press for that kind of coordinated response. There are severe inherent problems that you recognize better than anyone.

For one thing, I might add, is where would the congressional responsibility for that coordination lodge? Are you going to carve out areas that now are scattered among five or six defense committees and have one super committee take a unique responsibility for it? It is not just a question of reorganizing the executive branch.

Because we are dealing with preparing for standby capacity in large measure, I do not think we want to reinvent forces that would be waiting and performing no other function. So I think making the most effective use of existing public health resources, law enforcement resources, defense resources, and extending their training is important. So I do not favor having a single operating agency that has its own staff, troops, and so forth. I think we do have to find a way to coordinate existing programs.

Senator SPECTER. How do you do the coordination then?

Dr. LEDERBERG. Well, that is very, very difficult. Without having a centralized budget, you then have mostly powers of executive persuasion, let me call it—it is a phrase I just invented—to try to modulate the behavior of individual departments. I think there does need to be a very strong focus.

The only place in government that now exists that is remotely related to it is the one that is being used, is the National Security Council, which can call on the secretaries of other departments. It does not actually have a lot of executive clout because it does not have the budget, and I do not think we would want it to have the overall budget that would be involved.

So I think the strong backing of the President, the designation of an officer in the National Security Council, and then a more or less standing committee representing the different departments.

Senator SPECTER. Would you prefer someone else in the NSC to the Vice President?

Dr. LEDERBERG. I would say, unless you are going to greatly enlarge the Vice President's staff and create a new office in government which is not in the Constitution as the responsibilities of the Vice President, I would think the National Security Council would be the natural lodging. Perhaps the Vice President in some administrations——

Senator SPECTER. Let me turn to Dr. Henderson, because of the time limit, with the same question. You went through a long litany of what needs to be done. How would you make it happen? You have been in the White House. You have had extensive governmental experience. What is your best recommendation?

Dr. HENDERSON. Well, I think Dr. Lederberg has outlined a lot of the considerations certainly which have occurred to me as well. I think there is a lot that can be done from the White House perspective if indeed you have got strong leadership there and good support from the Office of the President, there is a lot that can be done—and goodwill. I think my feeling would be that this would be the logical locus.

Senator SPECTER. The Vice President?

Dr. HENDERSON. Well, I think that I have never seen the Vice President's office function in this way, but it is quite possible that that is the way to go. I think something in that area, however, is important.

Senator SPECTER. My yellow light is on.

Dr. Myers, I want to ask you a question about companies which will develop vaccines. You identify anthrax. Why is there only one producer of anthrax? How do we get moving on smallpox? Is there some way to stimulate the private sector to dig into this issue, or are they too worried about product liability issues?

Dr. MYERS. I am not sure how to stimulate the private sector, but it has been clear through the years that the private sector has had very little interest in working on defense vaccines. Probably that was more clearly established in the proposal by the DOD called the Joint Vaccine Acquisition Program, which was widely advertised. It concerned the development of some 18 biodefense vaccine products, a 5- to 10-year program with an estimated budget of $500 million. Not a single major manufacturer of pharmaceuticals applied.

I think besides the liability there is issues of profitability, there is issues of working with government. You call it getting ducks in a row here in Washington. In Michigan we call it herding cats. It really is not easy to work with government.

But as importantly as anything, why would a company risk its reputation when it has lines of vitamins that are worth half a billion dollars in sales a year, risk its reputation to be involved in defense vaccines when this issue is so uncertain and has drawn such criticism?

Senator SPECTER. Thank you very much, Dr. Myers.

Senator Rockefeller.

Senator ROCKEFELLER. Thank you.

Dr. Lederberg and Dr. Henderson, I think one thing is instructive when people do not take these potential threats seriously. I think it is the cult group Aum Shinrikyo which did its number in the Tokyo subway. What I think is less well known—but among the biological and chemical fraternity, they would know—is that that group tried, before they resorted to a chemical weapon, they tried very hard to use biological weapons, by introducing one into an exhaust system of automobiles which they were going to send around town, and the other with an aerosol approach from the top of a couple of buildings in downtown Tokyo.

They did not happen to work, and so they resorted to the chemical in the subway. But the fact that they did not work is really, I think, relatively unimportant, because making something like that work is, it seems to me, in the high realm of probability. The effect of that, it seems to me, would be extraordinary.

Dr. Lederberg, I guess this is the question I would want to ask you and Dr. Henderson: In the event of something of that sort, let us say, in smallpox, what would be the effect on a city the size of Seattle, let us say?

Dr. LEDERBERG. I am going to defer to Dr. Henderson on anything to do with smallpox.

Dr. HENDERSON. Yes, I had 11 years experience with smallpox and I know it only too well. This is a disease which is, I would say, one of the most horrible diseases that one could possibly ever see. I have certainly seen many in hospitals, with a 30-percent death rate, a disfiguring disease, painful, very miserable disease.

It does generate a great deal of fear, tremendous fear, tremendous anxiety. So when we have seen outbreaks occurring in Europe, one in Yugoslavia in 1972, the countries around simply close their borders immediately this was discovered and permitted nothing to go across borders. It is a reaction which is very extreme.

This is what I think we would expect to see in our own cities. We have worked out a scenario looking at this, trying to estimate what would happen, and we would foresee enormous difficulties. Particularly, this is a more communicable disease than we had even thought about ourselves until we worked through the European experience. December through April we were getting 10 cases for every 1 every 2 weeks. In other words, it is multiplying tenfold every 2 weeks.

With the movement of people and travel, one can assume that it is not going to be a nice little localized outbreak in one city, but it is very rapidly going to involve many cities, States. There is going to be a tremendous demand for vaccine. Whether it is warranted or not, the demand will be there, and I think we will use up vaccine very quickly.

At this point you begin to get a spiraling problem with more cases, suspect cases, more demand for vaccine, no vaccine available, nothing you can do. You can see the scenario that is unrolling. It does not take a lot of cases, I think, to do this.

I think we have got a lot to be concerned with smallpox.

Senator ROCKEFELLER. Is it not true that—although this does not have any enormous moment for the present—but it is interesting that back in the Middle Ages, in the case of the plague or smallpox or other things, that where walled cities were under attack and plagues broke out, as people died they threw the bodies over the walls so that they could land at the feet of the enemy, because even they understood that there was an incubation period or something, and afterwards it was something that could spread very rapidly?

Dr. LEDERBERG. That happened in 1436 in Kaffa, a Genoese fort on the coast of the Black Sea.

Senator ROCKEFELLER. Is there not some thought that in fact, that then carried on to the great plague, that that may have been one part of the genesis of the great plague?

Dr. HENDERSON. That was the beginning of it, that is correct.

I think with smallpox over the years we have been—countries have been much more concerned about this, more fearful of it, than any other disease. In fact, Britain until the early eighties maintained four hospitals on a standby basis, to be opened only if cases of smallpox came into Britain. Germany built two new ones in the 1960's simply for that purpose.

So that there is great concern about this disease, and in fact even a small outbreak I think would generate some real serious problems so far as civil disorder was concerned.

Senator ROCKEFELLER. I am always impressed when distinguished people are willing to come long distances for relatively short hearings. Thank you both, all of you, very much.

Senator SPECTER. Thank you very much, Dr. Lederberg, Dr. Henderson, Dr. Myers. We very much appreciate your being here.

CONCLUSION OF HEARING

That concludes our hearing, the subcommittee and Committee will stand in recess subject to the call of the Chair.

[Whereupon, at 10:57 a.m., Tuesday, March 16, the hearing was concluded, and the subcommittee and Committee were recessed, to reconvene subject to the call of the Chair.]

○

APEX
TP
9780160601941
76901945 – 24